Reality and Its Order

Werner Heisenberg
Reality and Its Order

Edited by Konrad Kleinknecht

Translated from German by M. B. Rumscheidt,
N. Lukens and I. Heisenberg

Author
Werner Heisenberg
München, Germany

Editor
Konrad Kleinknecht
Heisenberg-Gesellschaft
München, Bayern, Germany

Translated by
Martin B. Rumscheidt
Dover, NH, USA

Nancy Lukens
Dover, NH, USA

Irene Heisenberg
Durham, NH, USA

ISBN 978-3-030-25698-2   ISBN 978-3-030-25696-8   (eBook)
https://doi.org/10.1007/978-3-030-25696-8

Translation from the German language edition *"Ordnung der Wirklichkeit"* by Werner Heisenberg © Piper Verlag GmbH, 1994.

© Springer Nature Switzerland AG 2019

This work is subject to copyright. All rights are reserved by the Publisher, whether the whole or part of the material is concerned, specifically the rights of translation, reprinting, reuse of illustrations, recitation, broadcasting, reproduction on microfilms or in any other physical way, and transmission or information storage and retrieval, electronic adaptation, computer software, or by similar or dissimilar methodology now known or hereafter developed.

The use of general descriptive names, registered names, trademarks, service marks, etc. in this publication does not imply, even in the absence of a specific statement, that such names are exempt from the relevant protective laws and regulations and therefore free for general use.

The publisher, the authors and the editors are safe to assume that the advice and information in this book are believed to be true and accurate at the date of publication. Neither the publisher nor the authors or the editors give a warranty, expressed or implied, with respect to the material contained herein or for any errors or omissions that may have been made. The publisher remains neutral with regard to jurisdictional claims in published maps and institutional affiliations.

This Springer imprint is published by the registered company Springer Nature Switzerland AG
The registered company address is: Gewerbestrasse 11, 6330 Cham, Switzerland

Editor's Preface

Werner Heisenberg, Nobel Prize winner for his discovery of quantum mechanics and one of the eminent scientists of the twentieth century, wrote this essay during the war years 1941/42. Only relatives and reliable friends obtained a copy, but he did not think of publishing it. Considering some of the contents, this would have been very dangerous in the political situation. Therefore, the text is a sketch without references. In this essay, Heisenberg summarizes his philosophical thoughts about nature and about the question how man can know what reality is.

On July 10, 1941, Heisenberg wrote to his wife Elisabeth: "Towards evening I wrote on the private philosophy, and started the passage about the roses. I now write on these things with great enjoyment. Not always with a clear conscience, because, basically, I understand almost nothing of all these things. But since Bohr probably will not write down his thoughts, it is good that anyone who knows them is writing down what he makes of it. In Urfeld I could maybe seat myself at the little table in the bushes and also continue pursuing these thoughts".

After Heisenberg's death, the editors of his 'Collected Works' ("Gesammelte Werke") published this text under the title 'Ordnung der Wirklichkeit'. It appeared in 1989 with Piper publishers and included an introduction by Helmut Rechenberg. The Heisenberg Society issues here a new edition, for the first time in English, and with the addition of a commentary on the literary, musical, philosophical and historical background. The commentary is written by the science historian Ernst Peter Fischer. The citations [C1], [C2], etc. in Heisenberg's text alert the reader to the existence of a related comment.

The Heisenberg Society is grateful to the Heisenberg family for their permission to publish this new edition. In particular, Irene and Jochen Heisenberg have contributed in many ways to the genesis of this work, and our sincere thanks go to them. Likewise, we thank Max Rechenberg for allowing us to include the introduction by his father, Helmut Rechenberg, written in 1988. We also acknowledge the excellent work of the translators M. B. Rumscheidt and N. Lukens. For editing the book and her ever-friendly collaboration we thank Angela Lahee.

München, Germany
June 2019

Konrad Kleinknecht

Contents

Introduction   1
Helmut Rechenberg
    The Philosophically Interested Colleagues Among Heisenberg's
    Circle of Physicists   3
    On the Origin of the Philosophical Manuscript "Reality
    and Its Order"   8
    Notes on the Contents of the Essay and Conclusions   12
    References   17

Reality and Its Order   19
Werner Heisenberg
I   19
  1. The Diverse Areas of Reality   20
  2. Language   24
  3. Order   29

II   34
  1. The Domain of Reality in Goethe's View   34
  2. (Classical) Physics   39
    (a) Mechanics   39
    (b) Electricity and Magnetism   42
    (c) The Infinite   45
  3. Chemistry   51
    (a) Heat   52
    (b) The Laws of Chemistry   54

	(c)	The Boundaries of the Domains	59
	(d)	Chance	63
4.	Organic Life		66
	(a)	The Relation Between Biological and Physical-Chemical Laws of Nature	67
	(b)	The Structure of the Biological Domain	74
	(c)	The Unique Position of the Human Being	81
5.	Consciousness		83
	(a)	Consciousness and Biology	83
	(b)	Consciousness and Reality	86
6.	Symbol and Gestalt		89
	(a)	The Means of Communication	90
	(b)	Art	98
	(c)	Science	101
	(d)	The Symbols of the Human Communities	104
7.	The Creative Forces		107
	(a)	Religion	110
	(b)	Illumination	113
	(c)	The Great Parable	116

III   118

Commentary on Werner Heisenberg's "Reality and Its Order"   123
Ernst Peter Fischer
   Preamble to the Commentary   123
   The Comments   125
   References   147

Author and Editor

About the Author

Werner Heisenberg (born 1901 in Würzburg/Germany—died 1976 in München) is one of the leading scientists of the twentieth century, inventor of quantum mechanics and Nobel Prize Winner. Heisenberg studied physics with Arnold Sommerfeld in Munich and with Max Born in Göttingen and worked as an assistant to Niels Bohr in Copenhagen. On the island of Helgoland in 1925, he made the breakthrough to a theory of the atom, dubbed Quantum Mechanics. In 1927, he found that in the atomic world, there are limits to our knowledge, which he specified as the Uncertainty Relation. In 1933, he received the Nobel Prize as "creator of the theory of Quantum Mechanics". From 1945, he was director of the Max Planck Institute for Physics and president of the Humboldt Foundation.

About the Editor

Konrad Kleinknecht (born 1940 in Ravensburg) is professor of experimental Physics at the Johannes-Gutenberg University of Mainz and member of the excellence cluster "Universe" at the Ludwig-Maximilians-Universität München. He has worked at the universities of Heidelberg, Dortmund, Mainz and Munich, at the European Laboratory for Elementary Particle Physics CERN in Geneva/Switzerland, Caltech in Pasadena and Fermilab near Chicago and gave the Loeb lectures at Harvard. His work on elementary particle physics has been recognized by numerous awards.

Introduction

Helmut Rechenberg

Physicist Werner Heisenberg (1901–1976) is one of the great natural scientists who have given shape to how the world will view itself beyond the twentieth century. He succeeded in establishing the point of departure of today's quantum mechanics and made specific contributions to the effective description of atoms and molecules. His indeterminacy relations provided the key to the physical-epistemological interpretation of this new theory. Finally, he did decisive pioneer work in the expansion and coordination of quantum and relativity theory. Above all, he confronted problems of the innermost structure of matter. He was engaged, in other words, in what we today call nuclear and elementary particle physics.

In lectures and articles Heisenberg frequently took a position on questions that went beyond the narrower boundaries of his scholarly specialty. He particularly sought to make the findings of "modern physics," its epistemological foundations and philosophical conclusions accessible to a broader public. This led to individual publications as well as collections of articles with such titles as "*Die Einheit des naturwissenschaftlichen Weltbildes,*" (The Unity of the View of the World in Natural Science), or "*Wandlungen in den Grundlagen der Naturwissenschaften*" (Transformations in the Foundations of the Natural Sciences). Beyond that, Heisenberg wrote three extensive texts on philosophical questions dealing with the description of nature. These are his "Gifford Lectures" delivered during the winter term of 1955/56 and later published as a book entitled *Physik und Philosophie*, (Physics and Philosophy) 1958, 1959, his memoirs *Der Teil und das Ganze*, (The Part and the Whole) 1969, and the present extensive essay. It existed only as an untitled and undated manuscript before the publication of Heisenberg's *Gesammelte* Schriften, Collected Works. It is presented here for the first time as a separate publication.

We have titled the essay *Ordnung der Wirklichkeit*, (Reality and Its Order) in accordance with a characterizing remark by the author in the text itself. Written prior to the end of 1942, this is Heisenberg's earliest thorough and, on the other hand, thematically most encompassing statement he ever made on the philosophical and epistemological substance of the understanding modern physics has of the world. Here as never before, Heisenberg tries systematically to describe the whole of reality confronting the human being—from physical and chemical phenomena to biological systems up to the orders of society and the ideas of art and religion. Many of these questions are, indeed, touched upon again in later works or in the reminiscences of *Der Teil und das Ganze*, but they appear in *Ordnung der Wirklichkeit* in such an original and programmatic combination that we may describe this long essay as a kind of epistemological to all of Heisenberg's work.

Heisenberg's text is divided into three parts. Part I, an Introduction, outlines in three sub-sections the aforementioned "Areas of Reality," the "Language" used to describe them, and their "Order." The Main Part (II) begins with 1. introductory remarks on Goethe's poetic ordering of the areas of reality which had given Heisenberg the impetus for the essay. It then develops a six-point schema of reality and its order. The schema is built up from the lowest areas as follows: 2. Classical Physics, 3. Chemistry including quantum theory, 4. Organic Life, 5. Consciousness, 6. Symbol and Gestalt. Part III, is a Conclusion where the author comments on the political conditions of the time, giving perhaps the impression of a "Consolation of Philosophy" to his preoccupation with the order described.

It is not the task of these remarks to analyze the essay's multifaceted content; that is left to the reader. But a few references ought to be made that may help understand the text more easily and permit it to be placed into the tradition of similar writings, into its historical context and into Heisenberg's biography. Three questions are to be addressed; First: how does Heisenberg see his own place among his contemporary philosophizing physicists? Second: How and when did the present text come into being? And third: What conclusions relating to the author's special views may be drawn from the text?

The Philosophically Interested Colleagues Among Heisenberg's Circle of Physicists

The relation between physics and philosophy that had emerged from a common ground in ancient Greece had been badly affected, if not entirely dissolved, in Central Europe since the mid-nineteenth century. The exact natural sciences had energetically turned particularly against the speculative natural science of the Schelling School. Even though some significant pioneers of the new "speculation-free" physics, such as Hermann von Helmholtz or Ernst Mach, addressed important epistemological issues, physicists in general restricted themselves to their special tasks and, in so doing, deepened and expanded physical knowledge immensely. But the decisive transformations in the foundations of physics at the beginning of the twentieth century that quantum and relativity theory had brought about forced a discussion of its philosophical consequences. This was needed especially in light of the fact that early "classical physics" had found a firm place in later philosophical thought such as Newton's mechanics in Kant's Critique. Again, it was precisely those physicists who had substantially shaped the radical transformation, namely Albert Einstein and Max Planck, who were the first to contribute to the philosophical-epistemological discussion. This is not the place to address the extensive debates on the theory of relativity elicited by Einstein who was trained in Mach's epistemological methods, nor those on the foundations and conclusions of quantum theory, debates that continue to this day. It must suffice to recall some epistemological and philosophical questions that occupied physicists in Heisenberg's field and that emerged from the results of their work's results.

Among Heisenberg's physics teachers Arnold Sommerfeld hardly paid attention to philosophical problems; Max Born did so only quite late in his life. Niels Bohr (1885–1962) was a very different case. It was from him that young Heisenberg, characterized by his friend Wolfgang Pauli—and not only by him—as being "unphilosophical," "brought home a philosophical orientation of his thinking."[1] Bohr, the teacher, achieved success, as Pauli was later to confirm in a letter of 27 July, 1925 to Hendrik Kramers, "I also noticed with delight that Heisenberg learned a bit of philosophical thinking from Bohr in Copenhagen and now noticeably turns away from the purely formal."

---

[1] In a letter of 11 February, 1924, by Pauli to Bohr. The citations of letters by Pauli are taken from [1].

It is remarkable that Bohr did not really publicly address questions ranging beyond pure physics at all until about 1930. Pauli seems to have referred to the special way in which the Copenhagen physicist went about the problems of quantum physics, namely his precise and logically faultless discussion of physical phenomena and their foundations. It was that discussion Heisenberg came to know and appreciate during longer stays with Bohr before it appeared in the latter's lectures and writings for a public not specialized in physics. During the 1930s Bohr sought particularly to extend his "principle of complementarity," formulated first in 1927, from atomic physics to many other areas. This principle stated that certain phenomena permitted two wholly exclusive descriptions and viewing the two "complementary" methods of description alone yields a complete picture. Thus, he discusses chemical problems (1930), biological processes (1932, 1937, 1957, 1962), and the relation of physics to psychology (1938). He tried also to introduce the idea of complementarity into the study of human cultures (1938, 1954, 1960). Bohr's lectures and articles were collected in two volumes entitled <u>Atomphysik und menschliche Erkenntnis</u>, (Atomic Physics and Human Knowledge) (1958, 1966).

Heisenberg owes crucial insights to Bohr's epistemological-philosophical discussions. In his first publications for a general public he already follows closely in form and substance the train of thought of his teacher in atomic physics. On the occasion of Bohr's fiftieth birthday, he writes in particular:

> For the scientists who had the good fortune of having the chance to work for a time in Bohr's institute in Copenhagen another aspect of Bohr's work [besides physics per se] is almost more important. It is the creation of an intellectual center where the most diverse threads of modern natural science come together and enter into relation to the general substratum of all natural, physical and human sciences. The extraordinary personal influence Bohr had and still has on his students is rooted precisely in this unity of thought where every scientific question, just as life itself, is brought into relation to the same, unchangeable center.[2]

The center he speaks of is, of course, the principle of complementarity that came to occupy a central place in Heisenberg's thought.

Another founder of modern atomic physics, not one of Heisenberg's academic teachers yet influencing him increasingly through his writings

---

[2]Heisenberg [2]. Reprinted in Heisenberg [3]; cited hereafter as GS/CW with the appropriate vol. no. and date.

particularly after 1930, was Max Planck (1858–1947) the father of the quantum theory. It was especially what Planck wrote on the relation of physics to issues of philosophy, politics and religion that made its mark on Heisenberg. Only after he had turned 50 did Planck express himself on topics that went beyond the substance of physics. In 1908 he lectured in Leyden on *Die Einheit des physikalischen* Weltbilds, (The Unity of the Understanding of the World in Physics,) a polemic against the positivistic and anti-atomistic views of Ernst Mach. Others of Planck's lectures have revealing titles such as *Die Stellung der neueren Physik zur mechanistischen Weltanschauung* (1910) (The Position of Recent Physics Toward Mechanistic Interpretation of the World), *Dynamische und statistische Gesetzmässigkeit* (1914) (Dynamic and Static Regularity), *Kausalgesetz und Willensfreiheit* (1923) (The Law of Causality and the Free Will), *Positivismus und reale Aussenwelt* (1930) (Positivism and the Real Outer World), *Ursprung und Auswirkung wissenschaftlicher Ideen* (1933) (Source and Impact of the Ideas of Science), *Die Physik im Kampf um die Weltanschauung* (1935) (Physics in the Struggle for the Perception of the World), *Religion und Naturwissenschaft* (1937) (Religion and Natural Science), *Determinismus und Indeterminismus* (1938) (Determinism and Indeterminism), *Sinn und Grenzen der exakten Wissenschaften* (1941) (The Meaning and Limits of the Exact Sciences), *Warum kann Wissenschaft nicht populär sein?* (1942) (Why Can't Science be Popular?), *Wissenschaftliche Streitfragen* (1945) (Disputes and Issues in Science) and *Scheinprobleme der Wissenschaft* (1946) (Sham Problems of Science.) The very fact that this scholar, a man of integrity and respected world-wide, did not remain silent in spite of his personal rejection of the "Third Reich," in a time of great difficulty for science and scientists, but actually increased his activity as a public lecturer, gave strong intellectual and moral support to many colleagues in the field and to interested lay people.

Heisenberg's decision in 1933 to remain in Germany allowed him to move closer to Planck even though the latter's interpretation of quantum mechanics was contrary to Heisenberg's physical interpretation. In his review of Planck's anthology Wege zur physikalischen Erkenntnis (1933) (Pathways to Knowledge in Physics) Heisenberg concludes: "The overall impression evoked by Planck's lectures leads this reviewer to this summation: it is precisely Planck's religious-ethical perception of life that in the end determines his position vis-à-vis the epistemological situation of modern physics that permits him to walk a straight and almost too sure road even when at every turn of that road unfathomable chasms of epistemology threaten."[3] What

---
[3]Heisenberg [4], reprinted in GS/CW, CIV, p. 239.

Heisenberg meant by the "almost too sure road" was, above all, Planck's decisive defense of the strict validity of the law of causality.

Heisenberg's positive review met with Pauli's reproach. He wrote Heisenberg that he "noted with some displeasure certain phrases in the review of Planck's book," such as the admission that Planck's concept of "the reality of the outer world" was a valid one. Pauli implored Heisenberg: "May the spirit that hovers over Planck's scientific production and his personal life not gain all too much the upper hand in your publications and your life."[4] Pauli never forgave Planck for his polemics of 1908 against his god-father Mach. He believed that there were "qualities in Planck's activity" that he found "deeply, not at all superficially sloppy." He now felt that he had to criticize not only the scholarly but also the political Planck, who after the National Socialists' take-over, tried to keep some colleagues in Germany. Heisenberg did indeed agree partially with Pauli's objections to Planck's philosophy but not with his reproach of Planck's political and moral stance. A book review of 1935 concludes with these words: "Finally, Planck asserts with the entire solemnity of his being, that science, through its very nature, educates us in truthfulness. That makes him, beyond the domain of scholarly achievement, the spokesperson for German natural science. The most important and greatest task today is to guard that heritage."[5]

Heisenberg found support in Planck's political and human demeanor; he responded to what Planck dealt with in his lectures and articles. He often even adopted their titles despite the fact that his conclusions differed occasionally from those of his model. With Planck Heisenberg opposed the "unflinching positivists à la [Philipp] Frank," whereas his colleague at Göttingen, Pascual Jordan (1902–1980)—almost of the same age as Heisenberg—clearly represented the positivist method. In the thirties Jordan published a sizeable number of articles seeking to draw the philosophical consequences to be derived from quantum mechanics. The titles of his books signal the direction Jordan was pursuing in. Physikalisches Denken in der neuen Zeit (1935) (The Thinking of Physics in Recent Times), Die Physik und das Geheimnis des organischen Lebens (1945) (Physics and the Mystery of Organic Life), Eiweissmoleküle (1947) (Protein Molecules), Verdrängung und Komplementarität (1947) (Displacement and Complementarity), Atom und Weltall (1952) (The Atom and the Universe), Der gescheiterte Aufstand (1956) (The Failed Revolt). With his contributions after 1930 to biology,

---

[4]See note 1 above; vol. 2, p. 214.
[5]Heisenberg [5], reprinted in GS/CW, vol. CIV, p. 240.

Jordan crossed the boundaries of physics as a discipline and with them helped establish the Treffertheory of genetics. He clearly was perceived in his time as a pioneer of the new interdisciplinary biophysics.

Wolfgang Pauli (1900–1958) was also aligned with positivism, especially with the epistemological-critical method of Ernst Mach. He confessed in 1954, "So as to alert the philosophers, I would like to say that I do not belong to any philosophical school that has a name ending in 'ism,'" adding that he tended "to maintain a certain middle position between extreme orientations."[6] Pauli himself published little about more general philosophical problems of science. His collection Aufsätze und Vorträge über Physik und Erkenntnistheorie (1961) (Articles and Addresses on Physics and Epistemology) contains only five items on this matter. Among them are the important articles: *Phänomen und physikalische Realität* (1954) (Phenomenon and Physical Reality), *Naturwissenschaftliche und erkenntnistheoretische Aspekte der Idee des Unbewussten* (Natural Science and Epistemological Aspects of the Idea of the Unconscious) (1954, on the occasion of C. G. Jung's eightieth birthday) and *Die Wissenschaft und das abendländische Denken* (1955) (Science and Western Thought).

Heisenberg often profited significantly from the frank critique his friend offered to his philosophical writings. For example, Pauli caused Heisenberg to tighten a number of formulations in the essay *Der Begriff 'abgeschlossene Theorie' in der modernen Naturwissenschaft* (The Concept of 'Closed Theory' in Modern Science).[7] For his part, Heisenberg devoted an extensive essay to presenting Pauli's philosophical views.[8] There he refers specifically to two completely different sides in his friend's being and thinking. "The power of the fascination that emerges from Pauli's analysis of problems in physics came only partly from the clarity of his formulations, transparent to the last detail, but also partly from his constant contact with the realm of productive intellectual processes [in the subconscious] for which there is yet no rational formulation."[9]

Among Heisenberg's younger contemporaries, his pupil Carl Friedrich von Weizsäcker, the only one to be mentioned now, showed an interest quite early in philosophical questions. He had intended initially to study philosophy but Heisenberg persuaded him to study physics first as a foundation

---

[6]Pauli [6]. See esp. p. 93.
[7]It appeared first in Dialectica, vol. 2 (1948); reprinted in GS/CW CI, pp. 335–340.
[8]Heisenberg [7]; reprinted in GS/CW CIV, pp. 113–115.
[9]See note 8; p. 113. Pauli referred in particular to Jung's archetypes and occasionally used the symbolism of the alchemists.

for philosophy. After 1940 von Weizsäcker shifted the core of his work systematically to the epistemological analysis of the new theories. In one of his first major articles, on the relation of quantum mechanics to Kant's philosophy, he already sounded a central theme. Heisenberg had earlier become acquainted with Kant, primarily through the discussion on causality; through his pupil he now gained a thorough picture of critical philosophy which then entered in some places into his own lectures and writings.

One can say that since then physicists have been less willing to occupy themselves with questions of epistemology or the philosophical conclusions of their science. On the other hand, the discussions among non-physicists—often little grounded in accurate knowledge—have in no way abated. A continuation of the interaction between physics and philosophy in the future on a level comparable to that of the days of Planck, Einstein, Bohr, Schrödinger and Heisenberg would be a welcome dialogue.

On the Origin of the Philosophical Manuscript "Reality and Its Order"

We have Heisenberg's own words outlining when his extensive essay came into being. In a letter of 10 February, 1947, to F. Kraus he comments: "I wrote it down in the first years of the War." But the genesis of its content goes much farther back. We have reconstructed here some essential aspects of the story of its formation.

Immediately after his discovery of indeterminacy relations in 1927, Heisenberg began to reflect on basic consequences for the human understanding of nature arising from those relations. The first extant lecture on this matter was addressed, characteristically, to philosophers: *Erkenntnistheoretische Probleme in der modernen Physik* (1928) (Epistemological Problems in Modern Physics.) There he sketches the difficulties created by the results of the theory of relativity—the new perception of space and time—and of quantum-theory—complementarity and indeterminacy relations—for the classical epistemology formulated by Kant.[10] The relation of the law of causality and quantum mechanics occupied Heisenberg especially in the thirties, beginning with his programmatic lecture in 1930 at the conference of natural scientists held in Königsberg.[11]

---

[10] GS/CW CI, pp. 22–2.
[11] Heisenberg [8]; reprinted in GS/CW CI, pp. 29–39.

At other times he reported on the further developments in the idea of complementarity beyond atomic physics (in biology, psychology) that Niels Bohr in particular was presenting. He formulated his idea of "closed systems" in physics, that is, of theories that present with a certain degree of consistency and completeness specific areas of experience of physics.[12]

After the year 1933, Heisenberg gave increasingly more lectures apart from those at specialist conferences; this came about on account of the ideological attacks by influential people of the "Third Reich" against modern theories of physics and their proponents. Heisenberg played a prominent role in the defense of the physics disqualified as "Jewish." This dispute offered the opportunity to consolidate the epistemological foundations of physics and to present its philosophical consequences more clearly. In the first years of World War II, when the opposite side, the so-called "German Physics," propagated by Philipp Lenard, Johannes Stark and their followers, retreated, the dispute finally came to an end.[13]

The dark time from 1933 to 1937 was both personally and professionally dangerous for Heisenberg; visits to friends outside Germany, such as Niels Bohr in Copenhagen and Wolfgang Pauli in Zurich, brought light into that darkness as did the contacts with like-minded colleagues in Germany. A small circle of professors at the University of Leipzig that Heisenberg belonged to allowed relaxed and informative conversations on matters that went beyond the narrower area of their special disciplines. The circle was called "Coronella"; in addition to Heisenberg there was the art historian Theodor Hetzer, the Nordic linguist Konstantin Reichhardt, the historians Helmut Berve and Hermann Heimpel, the archeologist Bernhard Schweitzer and the classical philologists Friedrich Klingner and Wolfgang Schadewaldt. They met regularly in private sessions where they presented their lectures, followed by informal discussion. The lectures and addresses they gave in public either corresponded to those in their private circle or had had their "dress rehearsal" there. Later Heisenberg maintained his relationship with the members of "Coronella" over many years.

War began on 1 September, 1939; a few weeks later, Heisenberg was conscripted to the Ministry of Ordnance in Berlin. He was assigned to cooperate in the secret German uranium project and, as a first step, to ascertain whether splitting the uranium atom in a chain reaction could prove useful

---

[12]Cf. Heisenberg's address during the gathering of natural scientists in Hannover [9], reprinted in GS/CW CI, pp. 96–101.
[13]Cf. Heisenberg's essay *Die Bewertung der 'modernen theoretischen Physik'* of 1940, published in the Zeitschrift für die gesamte Naturwissenschaft, vol. 9 (1943), pp. 202–212.

for the production of energy or explosives. Heisenberg's foundational theoretical studies (December 1939, February 1940) allowed for the possibility of producing energy. Experimental measurements carried out at different institutes in the German Reich, including those by Robert and Klara Döpel with Heisenberg in Leipzig, showed that a "uranium machine" with natural uranium and heavy water as moderator would work. When this successful result became apparent, Heisenberg wrote "Coronella" colleague Heimpel. He thanked him for the copy of his book Deutsches Mittelalter (Medieval Germany) Heimpel had sent him, saying: "I very much liked what you wrote there about how the Middle Ages thought of their times; it is so contrary to our epoch. And it occurred to me at that moment that such a transformation might come about again in the near future. For we humans could one day perhaps realize that we do in fact possess the power utterly to destroy the earth, that, in other words, by our own doing we are capable to conjure up a 'doomsday' or something close to that. But it is probably a flight of imagination to think such thoughts." Heisenberg wrote this on 1 October, 1941.

> That "flight of imagination" – Heisenberg most likely had the atom-bomb in mind was closer to reality than he imagined. In Germany, though, physicists were spared from traveling the road to the bomb.

Heisenberg found little time to work on other problems of his discipline and those beyond it, given the intensive efforts on the uranium project during the first two years of the war that required constant travel between Leipzig and Berlin. But his position as consultant at the Kaiser Wilhelm Institute for Physics, where the ministry had lodged the highest administration of Germany's uranium research, allowed him to conduct colloquia there on various subjects. In the summer term of 1941, he led a biology and physics colloquium that included lectures by, among others, N. W. Timofe'eff-Ressovsky, Ernst Zimmer and Karl Friedrich Bonhoeffer, on radiation biology. Another colloquium that term was called "The Physics and Chemistry of Proteins" where, among others, Alfred Butenandt, Karl Wirtz, Heisenberg and Carl Friedrich von Weizsäcker gave lectures. Heisenberg also invited Pascual Jordan to those colloquia; he was at that time doing military service at the air force headquarters in Bremen-Neuenlanderfeld. Other colloquia, beginning in the winter term of 1941/42, dealt with cosmic radiation and the theory of elementary particles. Some of those lectures were revised and published in 1943 in a volume called Kosmische Strahlung (Cosmic Radiation).

Heisenberg's war-related and additional activities restricted his lecturing and public speaking to a certain extent. On one of his rare trips outside Germany during those first war-years, to Budapest, he delivered a lecture on 5 May, 1941, on *Die Goethe'sche und die Newton'sche Farbenlehre im Licht der modernen Physik* (Goethe's and Newton's Theory of Color in Light of Modern Physics).[14] It was not a new subject matter for him, for as Elisabeth Heisenberg recalls: "When we were married, my husband showed me pictures one day that he had carefully guarded in his desk. They were pictures of single flowers but that obviously had some inner connection. They had been painted by Leipzig artist Hildegard Kress. She had attended one of my husband's lectures in the Academy on Goethe's theory of color and was so inspired by it that she painted those pictures in accordance with the principles she heard outlined from that theory of color. Then she presented them to my husband; it must have been around 1935."[15]

Heisenberg's Budapest address and its subsequent publication met with fairly widespread resonance. It stimulated the author to attend more often to similar topics. In any case, he delivered at least four addresses during the next year and a half to a wider audience. The titles were: *Die Einheit des naturwissenschaftlichen Weltbildes* (Leipzig, 26 November, 1941) (The Unity of Natural Science's Understanding of the World), *Das physikalische Weltbild* (Physics' Understanding of the World) (an expansion of the previous lecture; place and date unknown), *100 Jahre Energiegesetz* (100 Years of the Law of Energy) (radio address of August 1942) and *Über das Weltbild der Naturwissenschaft* (Natural Science's Understanding of the World) (a further revision of the earlier Leipzig address, given most likely in Zurich, 27 November, 1942.)[16] Like the Budapest address, those addresses contain essential ideas that are then used or developed further in *Reality and Its Order*. For example, the last-cited address concludes, like the present philosophical manuscript, with a citation of the fairy tale about the duration of eternity.

Certain details in the manuscript of *Reality and Its Order* support the assumption that the author concluded the text sometime between the

---

[14]*Die Goethe'sche und die Newton'sche Farbenlehre im Lichte der modernen Physik.* Geist der Zeit. Wesen und Gestalt der Völker (Hochschule im Ausland) New Series, vol. 19 (1941), reprinted in GS/CW CI, pp. 146–160.

[15]The precise dating of Heisenberg's lecture on Goethe's theory of color could not be established at the Leipzig Academy. But the 1935 dating is supported to some extent by Heisenberg's review of Planck's publication noted in note 5 above. There Heisenberg makes explicit reference Goethe and Newton as investigators of nature with opposite approaches. In Planck's essay the comparison of the two men is of minor significance.

[16]The addresses named are republished in GS/CW CI, pp. 161–192, 193–201, 202–206, 207–215 resp.

middle and the end of 1942. In the section entitled "Symbol and Gestalt" ("Symbol and Form") Heisenberg cites biologist Konrad Lorenz' opinion that Kant's "a priori" forms of apperception are to be understood as "innate schemata" of the human being, equivalent to the instinctive acts of animals. Lorenz had dealt with that extensively in an article of the February 1942 issue of Die Naturwissenschaften, entitled *Induktive und teleologische Psychologie* (Inductive and Teleological Psychology[17]). Dating Heisenberg's manuscript to the year 1942 corresponds to Frau Heisenberg's reminiscence that her husband had written it in the company of his family. The family had not followed Heisenberg to Berlin in the autumn of 1942; they had moved a bit later to Urfeld in Bavaria.[18]

It is safe to locate the beginning of the manuscript's composition at the time of Heisenberg's Budapest address in May 1941, particularly since in its concluding section he cites the "organization of reality" spoken of in the manuscript. Certain passages of the Introduction lead one to think even that Heisenberg composed the Main Section during the summer vacation of 1941 and 1942 in Urfeld where, in 1939, the Heisenbergs had bought the house of painter Lovis Corinth. Frau Heisenberg confirms that he was highly concentrated as he wrote "with great engagement." In the fall of 1942, Heisenberg went to Berlin in order to prepare the planned major experiment for the uranium reactor. Hectic and restless times were, indeed, upon him again, leaving hardly any chance for philosophical "side-trips."

Notes on the Contents of the Essay and Conclusions

The physicist may well be surprised initially by the important role the author assigns to the great poet and "amateur scientist" Goethe in organizing his thought. Heisenberg admired Goethe from his early youth. Traveling to Heligoland in June 1925 to cure his hay fever and to discover quantum mechanics, he had Goethe's West-Östlicher Diwan in his luggage. Later Goethe's entire opus was to be found in Heisenberg's library. He read the great novels with the same care as he did all of Goethe's dramas, knew Faust

---

[17] Vol. 30 (1942), pp. 133–143. Heisenberg also cites Lorenz' idea in his Zurich lecture of October 1942; cf. note 16.

[18] "After 1942 the times became so turbulent and difficult that there was neither time nor energy for that kind of 'extra venture.'" (E. Heisenberg to H. Rechenberg, 12 April, 1984.).

and much of the poetry (by heart) as well as <u>Conversations with Eckermann</u>. The diverse versions of <u>Italienische Reise</u> (Italian Journey) were recited in the family. "Goethe accompanied him throughout his life," said Elisabeth Heisenberg. But where does the high esteem of Goethe the physicist come from? Had Heisenberg not made the following judgment about Goethe's famous feud with Newton, the hero of physics?

> It was clear to all who in recent times occupied themselves with Goethe's and Newton's theories of color that little insight is to be gained from examining which of the two is right or wrong. To be sure, one may decide about that in objective details and Newton's natural-scientific method will prevail over Goethe's intuitive powers in those few places where there is a real contradiction between them. But the two theories really address two different things so that one is left with the question how it is possible that such different objects can be attached to the concept of color.[19]

Heisenberg's idea concerning the different physical theories he called "closed systems" could now actually be applied to systems outside those of physics. That different "closed" physical theories could contradict one another in certain of their conclusions, even those like classical and quantum mechanics that dealt with the same subject matter, could now be taken as applying to other systems as well. It really made no sense at all to ask which theory was fundamentally correct; Goethe's and Newton's theories of color simply dealt with different "levels of description," that is, different "orders of reality."

It should be noted, of course, that Heisenberg adopted from Goethe the whole system of the "order of reality." One reason for this may be his predilection for the poet's work, another may be his basic conviction: "It is probably more accurate to assume that all truly great natural scientists were quite familiar with the language of poetry as well."[20] Thus the poet provides the order of the themes: "coincidental, mechanical, physical, chemical, organic psychic, ethical, religious, ingenious" and the natural scientist fills this structure with content.

Comparing Heisenberg's arguments with those of his fellow physicists and mentors, Planck and Bohr, one is struck by the significance he gives to language. Emphasizing the fundamental function of language would actually seem quite natural for the grandson of classical philologist Nikolaus

---

[19]Cf. note 14, p. 265.
[20]<u>Ibid.</u>, p. 267.

Wecklein and the son of August Heisenberg, scholar of Byzantine culture. What is remarkable is that even the young Heisenberg always highlighted the decisive value of language in writings of his discoveries in physics. To be sure, the concept of "language" has a much more wide-ranging meaning for him, as he develops in detail in this essay's section "On the Origin of the Philosophical Manuscript "Reality and Its Order"". Language also includes every depiction of reality such as those provided of physical phenomena by the mathematical formulae of theoretical physics. Thus, diverse languages are simply differing depictions either of the same area of reality—as in the case of the manifestations of atomic physics provided by the Göttingen quantum and wave-mechanics—or of different areas of reality. Heisenberg's point of departure here is important: "Every area of reality can ultimately be depicted in language. The abyss separating different areas cannot be bridged by logical deduction or consistent expansion of language." (See below, p. –).

Heisenberg's different "Orders of Reality" embrace both different "areas of reality" and different "idealizations of reality," for there is really only one world (reality) that confronts human beings.[21] The first area, "Classical Physics," (section "Classical Physics") comprises mechanics, electricity, magnetism and the theory of the space-time continuum, including the gravitation of matter, (i.e., the special and the general theory of relativity and their consequences for structuring the macro-world). The theory of heat and chemical reaction forms the second area of order: "Chemistry," the regularities of which are determined by the discontinuous atomic structure of matter, quantum mechanics and the partial intrusion of coincidence into physical description, (section "Chemistry"). Classical physics and chemistry as described here describe many, but not all, features of the third area: "Organic Life." In it, "the whole" likely represents more than the mere "sum total of all parts," seeing that we have both goal-oriented selection and evolution and inherited behaviors (section "Organic Life"). The fourth area: "Consciousness," is even less precisely definable. Consciousness sets human action apart from purely biological mechanism and forms the basis of higher intellectual and psychic processes (section "Consciousness"). Here as in the two final, highest areas: "Symbol and Form," (section "Symbol and Gestalt") and "The Creative Powers," (section "The Creative Forces"), Heisenberg developed a series of ideas that go far beyond what his predecessors and successors dealt with. Indeed, he did not shy away occasionally

---

[21] "Hence, classical physics may be defined other than by its individual areas by the fact that we may completely leave aside the method of understanding [that is, the method of observation] that instructs us about reality." (See below, –).

from contradicting his revered teacher Bohr, such as in section "Symbol and Gestalt" on the possible hereditary transmission of intellectual capacity.

The brief concluding Part III bears no title and diverges somewhat from the scope and style of the essay. Heisenberg adds a more individual and at the same time a political-moral dimension to the essentially objective orders of Part II and the introductory reflections of Part I which, too, were purely objective. Here the author discloses his personal thoughts on the problems of the day although he places them into broader and not time-specific contexts. Actually, the beginning of these reflections can already be detected in earlier sections. For example, in section "Symbol and Gestalt" Heisenberg speaks of the forces that hold human communities together, beginning with "the primitive feeling of the same race that is prevalent already in the animal kingdom." Then he names "common language" and finally refers to the yet stronger forces "that weld together peoples of different races and languages: a common faith" and, above all, "a common law." (See below, p. –). During the struggles of the Second World War, Germany, the Soviet Union and Japan based and defended their positions on an "ideology that is closely tied to the old ideology of national states" whereas the Anglo-Saxons based theirs on "the rule of common law and the wellbeing arising from it." (See below, p. –). The author in no way conceals his personal view on the movements fighting against one another at that time; e.g. in section "The Creative Forces" he speaks of National Socialism as a "peculiar this-worldly religion." By contrast, he characterizes the bonds that unite the Anglo-Saxon world in these words: "This other association sets out from the experiences of the first great minds of nascent modernity who, in addition to Christian reality drawn from revelation, discovered that other objective reality that set out on its triumphal march in the rise of modern natural science." (See below, p. –). During and after the War, especially outside Germany, people occasionally made conjectures about Werner Heisenberg's political stance, primarily about how he supported the actions of the National Socialist government. While some conjectures, among them the claim that he wanted to build the atomic bomb for Hitler, have been conclusively refuted by the facts, his remarks from *Reality and Its Order* cited above show that the sympathies of Heisenberg the physicist were by no means with that "peculiar this-worldly religion" but naturally with the "objective reality discovered by the first great minds of nascent modernity." He speaks even more plainly in the concluding comments (Part III) when he evokes the duty to "rebuild what has been destroyed and put our trust in other people, reaching across the din of passion." (See below, p. –). Furthermore, he writes "The few for whom the world is still luminescent" now need to take over the leadership, for "We have to keep reminding ourselves that it is more important to act humanely towards one another than to fulfill any professional duties, national duties or political duties." (See below, p. –).

Heisenberg tried to live by those maxims even in the final years of the war; his actions in Germany and elsewhere, such as his visits to countries occupied by Germany, confirm this. And there is yet another political idea to be gleaned from the Conclusion of the essay. In it, the author ponders the future shape of the world in which "science will play an even more important role than before." He assigns the greatest significance to pure science and says: "As long as the central realm of science remains untouched, it is unlikely that all that great a danger will arise from the fact that we control the forces of nature to a far greater degree than we used to." (See below, p. –). We know now that this hope of Heisenberg's was not fulfilled, particularly as in the context of the war that "greater level of control of nature's forces" was achieved. To this day, human beings exploit the new power created by natural science according to the conventional principles of base political morality. Yet, Heisenberg always was the optimist; he believed that "the ability of humans to understand [he also meant: to learn] is unlimited." (See below, p. –).

The attentive reader will readily see why this philosophical essay could not be published before the demise of National Socialist Germany. The size of the essay and its "too personal" nature caused the author to reject publication even after the war. On the one hand, he tightened and expanded some sections, especially those on physics and chemistry, sections "Classical Physics" and "Chemistry", for use in later lectures and articles.[22] On the other hand, certain statements, such as in the section on organic life (section "Organic Life"), had to be qualified in light of new insights in microbiology. But now, many years after its composition, those time-specific and personal elements recede in face of the historical significance of the document. Despite all limitations of and objections to it, *Reality and Its Order* is the great project of a Weltanschauung and represents at the same time a philosophical-epistemological testament and an honest testimony of the great physicist Werner Heisenberg.

The philosophical manuscript exists in one original script and two type-written copies which show minor variations.[23] The present version is

---

[22] Cf. for example the essay on "closed theories" named in note 7, or his Gifford Lectures <u>Physik und Philosophie</u>.

[23] One of the type-written copies was most likely written later, incorporating corrections most of which had been written into the hand-written original.

based upon Heisenberg's original script which includes several corrections not contained in the type-written copies. I thank Frau Elisabeth Heisenberg very much for her care in helping clarify the question of the text's origin its context. I am grateful to Professors Klaus Gottstein, Hermann Heimpel, Friedrich Hund and Carl Friedrich von Weizsäcker and Dr. Gerald Wiemers for their helpful recommendations, Dr. Walter Blum and Frau Barbara Blum and last but not least Mr. Ulrich Wank for their critical reading of the text.

München, July 1988 Helmut Rechenberg

References

1. Pauli W (1979 and 1985) Wissenschaftlicher Briefwechsel/Scientific Correspondence, vols 1 and 2. Springer, Berlin-Heidelberg-New York-Tokyo
2. Heisenberg W (1935) Niels Bohr zum fünfzigsten Geburtstage am 7. Oktober 1935. Naturwissenschaften 23:679
3. Heisenberg W (1986) Gesammelte Schriften/Collected Works, vol CIV. R. Piper Verlag, Munich-Zurich, p 41
4. Heisenberg W (1935) Review of Wege zur physikalischen Erkenntnis. Naturwissenschaften 21:608
5. Heisenberg W (1935) Review of Planck's Physik im Kampf um die Weltanschauung. Naturwissenschaften 23:321
6. Pauli W (1961) Phänomen und physikalische Realität. In: Aufsätze und Vorträge über Physik und Erkenntnistheorie. Fr. Vieweg & Sohn, Braunschweig, pp 93–101
7. Heisenberg W (1960) Wolfgang Paulis philosophische Auffassungen (Wolfgang Pauli's Philosophical Views). Naturwissenschaften 46
8. Heisenberg W (1931) Kausalgesetze und Quantenmechanik. Erkenntnis zugleich? Annalen der Philosophie 2:172–182
9. Heisenberg W (1934) Wandlungen der Grundlagen der exakten Naturwissenschaft in jüngster Zeit. Angewandte Chemie 47(1934):697–702

Reality and Its Order

Werner Heisenberg

I

Dedicating one's life to the task of exploring particular relations within nature automatically leads one to ask again and again how those particular relations arrange themselves harmonically into the whole, the way life, or the world present itself to us. To be sure, searching for individual natural laws often becomes an infinitely exciting game [C1]. The more convinced you are you have mastered the rules of nature, the happier such a discovery makes you. However, in the course of a lifetime even the most multifaceted and most artfully constructed game would be void of substance were it not set in relation to the universal. Hence, the mind ever and again circles around the problem of what holds that whole together which we call 'world' or 'life' (depending on whether we see ourselves excluded or included in it), and where in this whole those special inter-connections are to be found which we spend a great deal of our life's work trying to discover. This question is in turn connected to another, broader task.

Whenever a new, fundamental insight enters human consciousness at a particular moment of intellectual life, one must re-examine the question of what reality truly is. In the history of humankind, various epochs stand out in which the structure of reality has undergone clear changes. The question may be left open for now whether the basis for that structural change lay in a new insight, or whether that insight was only made possible by a change in the structure of reality itself. But we do sense a meaningful connection when we learn that three seemingly completely independent, yet inherently related, events [C2] took place in close temporal proximity roughly at the outset of the modern era: Columbus' first voyage to America, Luther's and Zwingli's debate whether in the Eucharist the bread <u>is</u> the body of Christ or <u>represents</u> it, and Copernicus' discoveries.

There are many indications that a profound change in reality is in the making in our time as well. The turbulent and fruitful years [C3] after World War I blew the first gusts of a new intellectual air into our only seemingly safe world and no one knows what will be "real" for human beings after the wars that have now begun. It is hardly a coincidence that in the natural sciences too, the understanding of reality too has become fundamentally transformed in the last few decades. Even if we do not yet fathom the connection between this transformation and those larger scale changes, the understanding of these special developments in natural science may at some later time prepare the way for a general intellectual development. Our era has, therefore, been given the task of comprehending the general features of modern natural science as the natural consequence of a specific relation to reality. The subject of this essay is that relation to reality although it can once again only be an expression of a particular time and its hopes.

1. The Diverse Areas of Reality

Today, we are no longer taught that the world around us is simple and undivided, a garden we walk through from birth to death, planted for us for pleasure or hardship. Whether nudged by science or forced by the storms of time, we ponder or doubt the fact that in our own lifetime reality has changed several times already—not only gradually as if on a walk through the countryside, but suddenly and unexpectedly and that these changes have perhaps caused deep disquiet in our consciousness or endangered the harmonious unity of our life.

Childhood memories reach back into a world with narrow limits of space and time, a world where "meaning" and "being" where not yet separated and where we could shape reality with magic powers according to our wishes and imagination. What was it like then? The thread from mother's sewing basket, laid out on the floor, became the high wire of the acrobat who showed his amazing feats at the fairgrounds last Sunday and I am the acrobat.

A piece of wood is the horse carrying me as the rider. It really is the horse; the material properties of the stick are only appearance. Over the years the world expands in space and time, the magic power to transform wanes and, forced by various experiences, we ourselves make room for the laws of matter in reality. But this reality is still the simple continuation of the childlike world we had formed.

And then another day comes back into memory. One morning, as so often before, the child climbs on the swing in the orchard at home and looks across the meadows down to the river and the hills on the other shore. Everything is as before. But suddenly the church steeple on the other side of the bridge begins to gleam in the sunlight. The brilliant light spreads over the bridge's columns and the poplars in the sloping meadow, climbing along the turns in the path through the field up to the great timber yard, from there to the beech grove on the hilltop until, as if by a stroke of magic, the whole world has been transformed. For the first time, even though ever so briefly, the child enters the new area of reality where, in its holiest place, love later comes to dwell. It will be a few more years until the world of childhood will have completely faded from view but between the reality that surrounds the child and that later one there is no gradual transition. From no other string instrument can ring out the sound of that silver string [C4] of which Gottfried Keller sang.

A sudden and uncanny change of reality can also break into the years of active productivity when new experiences hardly change the grownup's world. Too easily, for instance, we weave into our lives a guiding idea, a wish that soon seems to be the only meaning of this life. All good energies develop around this one wish and the belief that it can be fulfilled appears to be the very source of life. And then it can happen that fate suddenly shatters the foundations of that wish, showing once and for all that it cannot be fulfilled. In that moment, the world can change in the most uncanny way. People and things that spoke to us vitally remain mute and look rigid [C5] and unreal. Where once a meaningful coherence held our life together, there is now a rigid law that decides only according to cause and effect without concern for connections on a higher plane. In earlier times people used to say that God could forsake a human being. But perhaps in our time there are many for whom the world puts on a gray and rigid face.

It has often been remarked that in the different epochs of humankind's development, reality looked different as well. In their youthful phases, nations seem to possess a similar magic, transformative power similar to that we remember from our own childhood. In early classical Greece, the Greek people found themselves in a world inhabited by gods and demons;

countless trails led from the present to the earlier mythical past. Pan was a tangible presence in the solitude of the forests and the god could be present at worship in ways we can no longer reproduce.

History shows that in subsequent ages this power to live in non-material contexts diminished. The spreading influence of natural science and technology in the late Hellenistic period clearly shows how the way things relate to each other in the material world according to the laws of nature assumes more and more power in reality. But then the emergence of Christianity represented a sudden, unmediated transformation of reality. We know that this change caused the individual human being affected by it most profound inner upheavals. Augustine's <u>Confessions</u> [C6] for example are a moving document illustrating of the utter break conversion can bring about in the course of a life.

One could continue ad infinitum to name and elaborate upon the examples of such fundamental transformations of reality in history or in individual human lives. So it is probably best to accept for the time being that quite diverse connections can determine our life. If the word "reality" means nothing other than the totality of the connections that pervade and carry our life, then it is probably true that there must exist very different areas or layers of reality.

In this context, it should perhaps also be pointed out that the world where other organisms of our earth live differs tremendously from our world. What that means is that we can only indirectly draw conclusions about the world of organisms from the utterly different external conditions of their life, a world, that is beyond our direct apprehension. Uexküll's studies of the animal world [C7] come to mind here; he reconstructed the environment of the respective species from the anatomical structure of each organism and from the laws of physics. And yet all one has to go on in these studies are the external physical conditions of life. If one calls to mind how little can be concluded from these external conditions for the various areas of human life, one can vaguely imagine the wealth of possibilities that might unfold beyond these physical conditions.

One might easily argue, of course, when we speak this way about diverse areas of reality or even of different realities, that we are still dealing with only <u>one, undivided</u> reality that just looks different to different human beings or under different conditions. The differences, the objection continues, arise simply from the physical or mental tools aiding the living organism to enter into a relation with the world that operates according to unchangeable laws. Nothing can really be said in opposition to that conviction of the unity of the world if it is stated in the general form that ultimately, we simply wish to comprehend the whole world as a <u>single, coherent</u> nexus of related phenomena. But in the perception of the great age of natural science [C8] that came to a close at the beginning of the twentieth century, the view of the unity of the world was combined with another view, namely that this unity manifests itself directly through the course of the external material world governed by strict laws. That connection running its course in space and time was objective and clearly binding for all beings, without exception, whether they were living organisms or dead matter. This connection seemed to be the truly "real" world which reflected itself in the consciousness of living beings—sometimes distorted or dimly—as in a mirror. This understanding could claim in its favor that the processes of the intellect too are always somehow connected to material processes so that—since one could not doubt that material processes follow the laws of nature—it too was perhaps conditioned and prescribed by the material processes. And even when the obvious fact was emphasized then, that the mind's working differs qualitatively in every way from workings of the material process, it still seemed that a subjective intellectual world stood in opposition to the objective material world. The fact that the physical world follows the predictable laws of nature came to be seen at least as the solid skeleton that carries the structure of the world.

However, it was in this very question that the scientific exploration of nature during the last decades compelled a change of perception. For us the predictable course of natural phenomena in space and time is no longer the firm skeleton of the world but only one nexus among others that becomes separated from the web of relations that we call the world by the way we examine it, by the questions we pose to nature. This view resulted from the insight into the laws of nature, gained with the advances of natural science, according to which phenomena can no longer be reduced to processes in space and time.

That imposes anew the task of ordering, understanding and determining the diverse connections or "areas of reality" in their mutual relations. They need to be related anew to the separation of the world into an "objective" and a "subjective" reality. The boundaries of those areas have to be determined and it must be discovered how the one conditions the other. Thus, finally, we must persist until we gain an understanding of reality that comprehends the diverse coherent relationships as part of a single meaningfully ordered world.

Of course, it is not only as a result of recent scientific developments that reality came to be described as a web of diverse relations. On the contrary, what we have here is the renewed consideration of ancient, often explored chains of ideas. The justification to repeat what had been said often before is based only in the fact that the development of natural science in the past decades has thrown a unique new light on this view.

Perhaps this development justifies the hope that it must be possible to determine more accurately than before the mutual relations of the diverse areas of reality. It is likely that much of the confusion in the thinking about reality results from the fact that every single thing participates simultaneously in diverse webs of connection, just as every word relates simultaneously to diverse contexts. We need proof that a clear separation is at all possible, given these circumstances. Only an example that can demonstrate with mathematical clarity the mutual relations of two areas of reality will convince the reader that it is possible to order the various layers of reality clearly and to depict their boundaries.

2. Language

Whoever undertakes to examine reality in this manner needs language, the form in which human ideas, as in every ordered reflection, can be shaped and communicated. But already in this first step the examination stands at the edge of where all human insight occurs. Is it even possible at all to express something very specific by means of language? This question is not meant to imply that, yes, there are completely clear ideas but that language cannot always express them. Rather, the question refers to that unavoidable element of indeterminateness [C9], to that peculiar "hovering" in thought and speech that philosophers have described so strikingly.

How does the child learn language? [C10] In that curious interplay of receptive and interactive learning and acting on one's own about one year into life, the child tries to produce sounds and imitate ones he or she has heard. Sometimes a word, such as "ball," is uttered successfully. The adult's approval and the success caused by the pronunciation of the word must turn this sound into a kind of magic formula the range of which is then unconsciously measured by frequent repetition. After a short time, every toy the child wishes the adults to fetch and perhaps every wish it has of the grownups is called "ball." Only when the desired result fails to come about or the adults contradicting the child's wish is the child taught in the course of time—unconsciously—that the magic formula cannot be misused and gradually the word's range of use finds its limits. Only in the course of years does the sense of the word develop that corresponds to its ordinary use of language. The boundaries between the things that may be called "ball" and those to which that word does not apply are never drawn definitely.

Generally speaking then, the range of a word's usage is not sharply delimited. But there are other, more important reasons for the indeterminateness of language. What has to be emphasized at the outset is that a word's meaning may depend to a large extent on the context in which it is used. Closer inspection shows that there are no isolated concepts aligned with words {aligned with them} in such a way as to structure an idea into a sentence the way one lays bricks. Rather, every thought constitutes an indivisible unity and every concept contained in it obtains its particular nuance in this thought. A poet can express ideas that can no longer be spoken in ordinary language simply because the words take on a new meaning through their context, through the reverberation of other ideas or the poetic form of the sentence. The content of a poem cannot be reproduced in prose.

Furthermore, every concept can refer to quite different webs of connections all of which are mutually related by means of this concept. Take the word "color." It can denote the content of our consciousness or the property of a body such as a flower; it can be applied to the chemicals used for dying fabric. "Color" can refer to a physically objectifiable reality characterized by a wave length. Finally, in a transposed and general sense, it can refer to finer nuances of different qualities. But it may also mean nothing more than the word-symbol "color" itself, just as a symbol that only "signifies" structures of reality. In a certain sense, concepts are the conspicuous points at which the different layers of reality are interwoven. Whenever the question is raised about the ordered connections of reality, those connections are found in each case within a distinct layer of reality; the concept "layer of reality" can hardly be explained otherwise (One can speak of the effect one layer has upon another only when using a very general meaning of "effect"). On the other hand, each different layer is connected in the ideas and their concomitant words that as matter of course simultaneously refer to many connections.

Despite the ambiguity and indeterminateness of concepts, language is well suited—and was created—for somehow "representing" or "depicting" the conditions of reality or the ideas about them. This depiction cannot be complete or accurate. What it can do, however—to use a somewhat imprecise expression—is contain the "essential." What is meant by this is that in every re-presentation we direct our attention in the first instance to certain features that we then designate as "essential." For example, just as the human eye is capable of precise observation only in a small area of the retina and is always unconsciously directed toward the most important aspect of the image, similarly, human thinking focuses on a specific, small segment of subject matter which then enters into the clearest light of consciousness while the remaining content of the idea appears only in obscure semi-darkness. This "essential" aspect of an idea is what language can represent.

Nevertheless the representation of facts in language can take two different forms. One is "static" and the other "dynamic." The two may be distinguished but not sharply separated. On the one hand, language can, by continuously refining its concepts, try to achieve an increasingly accurate depiction of the same intended subject. Refining a concept means determining to the last detail the relations that exist between concepts, e.g., by tracing specialized concepts back to more general ones, or else by the ad hoc coordination of concepts with the contents of very specific experiences. Examples of such refining of concepts can be found in the languages of the sciences such as jurisprudence or in the description of nature in mathematical terms. In the end, one can establish a rigid schema of rules governing the connections between concepts, and then again between concepts and the contents of experiences, allowing an unambiguous determination of whether a sentence using this conceptual system is "correct" or "incorrect." But the question of how accurately this system of concepts represents the aspect of reality under consideration is measured only by its success. A complete and exact depiction of reality can never be achieved. But if the system of concepts in question proves itself to be successful, it is permissible to speak of an exact depiction of the "essential part" of the condition at hand. For the only thing firmly established by this is on what aspects our attention is to be directed. The most famous example of such a system of concepts is Newton's Mechanics, which, in a generalized sense, may be viewed simply as a component of the language of natural science. By definitions and axioms, language establishes firmly how concepts such as "mass," "force," "velocity," "acceleration," etc. are to be used and connected with one another. For purposes of ordering and describing mechanical processes, this system has proven itself so completely reliable that we can hardly doubt that Newton's mechanics accurately describes the segment of reality that can be spoken of with those words "mass," etc.

Of course, the very same refining of language that allows one to determine whether a particular sentence is "correct" or "incorrect," goes hand in hand in many instances with an impoverishment of the concepts occurring in that language. Unlike the words of ordinary language, the vocabulary of this kind of technical language refers to very specific context areas. That means that the part of reality that is depicted in technical language and that is "essential" for the scientific position adopted may seem unimportant in light of different perspectives. What was called above a "static" description of a part of reality is, therefore, unavoidably linked with a renunciation of grave consequence. What is sacrificed in "static" description is that infinitely complex association among words and concepts without which we would lack any sense at all that we have understood anything of the infinite abundance of reality.

There is another way of representing reality that can be set against the "static" one; we may call it "dynamic" representation. It is made possible by those infinitely complex associations among words. An idea expressed in this way is not meant to be as faithful a representation of reality as possible but to be the seed for further series of ideas. The issue is not the accuracy but the fruitfulness of concepts. As a result of the complex associations among words, new ideas attach themselves to one idea, further new ideas emerge from these, until finally in hindsight, a faithful depiction of an area of reality under consideration takes shape from the abundance of substance in the space the ideas have traversed and measured. This sort of representation is based in the vitality of the word itself. Here, a sentence can, generally speaking, not be "right" or "wrong." But one may call a sentence "true" when it fruitfully leads to an abundance of other ideas. The opposite of a "right" sentence is a "false" one. But the opposite of a "true" sentence will often be another "true" one. The most famous systematic formulation of this "dynamic" representation of reality is Hegel's dialectics [C11].

The domain of "static" thinking <u>explains</u>, since, after all, clarity is the goal of this form of thinking. The "dynamic" domain <u>interprets</u> as it searches for the infinitely complex associations among other domains of reality to which we can point.

A comparison can help make the characteristic difference between the two methods of thinking more comprehensible. Wishing to become very familiar with a certain landscape, one may rise above the particular territory in an aircraft and have precise optical instruments draw cartographic images of the land which can then be microscopically analyzed to the minutest detail. Such maps contain an accurate and complete picture of the landscape. But the researcher may also journey throughout the land, crisscrossing the area that is of interest, live there and, stimulated by every new observation to further exploits, discern ever new aspects of the nature of the land. This way, too, one will also get to know the countryside very well in the course of time. The image gained in the second way will not be accurate in the same way as the surveyor's image but, on the other hand, it does have features that are missing in the photograph even if the latter is accurate and in a certain sense complete.

Poetry, too, seeks to communicate knowledge of reality. Its presentation always has the dynamic character just described. Yet it goes even further than using the infinite associations inherent in the content of all concepts, in that in poetry words are woven into a formal complex that derives from the rhythm, meter and the entire cohesive form of language. Poetry, therefore, shares with the consummate forms of the representational type called "static" this gathering of concepts into a formal and, in a very general sense, "mathematical" arrangement. In a certain way, poetry stands at the place where the extremes touch each other: on one side, thinking purely in terms of content, fully exploiting the vitality of the word and, on the other, the linking of concepts in a strictly mathematical schema.

Generally speaking, every attempt to speak about reality will have both "static" and "dynamic" features at one and the same time. Clear, purely static thinking is in danger of deteriorating into form without content. Dynamic thinking can become vague and incomprehensible.

It is true that what exact natural science consistently strives for are closed systems of concepts and axioms that accurately reflect the perceived area of reality. But the course of research that, proceeding from familiar systems of concepts, seeks to organize a new area of experience, cannot proceed on the paths proscribed by chains of logical reasoning. The abyss between the already familiar conceptual systems and a new system may be leaped over by intuitive thinking but cannot be bridged by formal reasoning.

When we cross over from a clearly understood, already scientifically ordered domain of reality into a new one, we are once again in the situation of the child who has to learn simultaneously to think and to speak, who cannot speak yet since ideas that can be expressed are foreign to children, and who cannot think yet since children lack the concepts with which to order and connect ideas.

Even though the above makes apparent the narrow limits that are imposed on every specific scientific description of reality, there is, on the other hand, no reason to assume that the ability of humans eventually to understand particular areas of reality is limited in principle. On the contrary, this human ability to understand, to find one's way in reality, appears in every way to be unlimited. Just as children, seemingly without effort, learn to know and to understand the particular world into which birth has placed them—whatever the adults' language, actions or demands may be—researchers quite generally will relate to every area of reality they can experience in such a way that it can be called understanding, even if one can say only afterward what that word "understanding" means. Even though our thinking in a certain way always hovers over a bottomless depth, given that we can never advance step by step from the firm ground of clear concepts into the unknown territory, in the end it will nevertheless do justice to every new experience, to every accessible domain of the world. Our thought processes will always develop a language suitable to the envisaged domain of reality that precisely reflects how things are in this domain.

However far thinking may advance, of course, the feeling will always remain nonetheless, that beyond what has been researched there are other phenomena that defy formulation in language. Every time when there is an understanding of a new reality, their sphere of validity appears to be pushed yet one more step into an impenetrable darkness that lies behind the ideas language is able to express. This feeling determines the direction of our thinking, but part of the essence of thinking is that the complex relationship it seeks to explore cannot be contained in words.

Perhaps we can summarize what these last paragraphs sought to express as follows: Every domain of reality can finally be depicted in language. The abyss that separates different domains cannot be bridged by logical reasoning or coherent linear development of language.

The ability of human beings to understand is without limit. About the ultimate things we cannot speak [C12].

3. Order

In every age, people have tried to subject our knowledge of reality to some general order. Since the development of natural science in the modern age, and most likely influenced by its example, such attempts have usually started by asserting that there are certain insights the accuracy of which is beyond doubt. Such knowledge then becomes the starting point of a system in which one attempts, beginning from one such insight, to proceed step by step to discover other similarly certain insights into reality, until one begins to recognize in the process a general order of reality of everything knowable.

Of course, at different times these starting points have been radically different. Thus it was natural for the approach of natural science during the last hundred years, for example, to take observation through the senses as the starting point of such reflection. Presupposed was that immediate observation, the accuracy of which could be verified by other people, led to an unquestionable knowledge of reality. This view is in stark contrast to the understanding of earlier centuries, which points out, of all things, the deceptiveness of sense perception and saw the starting point for knowledge in "the pure ideas of the soul [C13] that goes into seclusion and returns to itself." To recall Malebranche p. 28:

"The human being who judges the things of the world solely by the senses ... will find himself in the most grievous state in the world being extremely far removed from the truth and happiness. But when someone else makes judgments about objects solely by means of the pure ideas of his soul ... then it is impossible for that person to fall into error."

Various systems consider mathematical truth to be the model of undeniable knowledge. Indeed, one cannot doubt mathematical statements that can be proven in terms of the axioms of the relevant area of mathematics. But it has often been pointed out that these are "analytical judgments," i.e. statements that derive from a process of unambiguous deduction from established presuppositions and foundational definitions. However, it was said, such statements cannot say anything about reality since no process of deduction can demonstrate that its presuppositions and definitions faithfully depict reality. Thus, mathematical truths cannot be used as a starting point for an order of reality.

Nonetheless, they can play a decisive role in every such process of ordering reality. Precisely because mathematical statements actually represent a form or order disconnected from all content, then conversely, every order can be depicted in mathematical forms and even more readily the more perfected it is. The appearance of mathematical forms in every comprehended domain of reality has stimulated the reflection of humans from early on. The Pythagoreans' studies [C14] of the rational relations of harmonic vibrations of musical strings, Plato's ideas about symmetrical bodies [C15], testify to the significance assigned to the mathematical form in the understanding of nature. Exact natural science since Newton is based on the silent presupposition that it must always be possible to order the areas of nature accessible to our experience according to strict laws that can be expressed mathematically. Even if one carefully analyzes other representations of reality as well, such as music or the creative arts, which are far removed from the natural sciences, they will reveal internal orders that are very closely related to mathematical laws. Those orders can be as clearly discerned as in a Bach fugue, for example, or in a symmetrical ribbon ornament; they might be noticed initially through a unique balanced quality, through the immediately evident beauty of a melodic line [C16] such as that of the famous sub-theme in the first movement of Beethoven's D major violin concerto. Closer examination always shows simple mathematical symmetries comparable to those mathematics deals with in group theory. Thus, mathematics is order par excellence, in its purest form, freed from all content.

Hence, in terms of content, mathematics cannot be the starting point for an order of reality. Generally speaking, it appears quite improbable to our age's scientific consciousness that an order of reality could begin with an understanding that is beyond doubt and then, on this basis, proceed step by step to encompass every domain of the world. It seems to us, despite Kant, that indubitable statements are [C17] always analytical and thus say nothing about reality. And synthetic statements cannot be seen as binding for all time even if they are *a priori*. The history of our views of space and time shows that even forms of apperception that precede experience and therefore must be called *a priori* are not necessarily substantive components of closed theories of space and time. Biologists have pointed out that *a priori* forms of apperception may perhaps be understood as "innate schemas" that as such are subject to the process of selection and are changeable over the course of millennia. And even if there were such a thing as unquestionable knowledge and it did not encompass all of reality in one fell swoop, no road could lead from one domain of reality we believe to know, to another, new one.

Thus, something other than a sure certainty of knowledge has to exist at the beginning of an order of reality and, as history teaches, that other something results from a free decision more likely to be taken by larger human communities or by humankind as a whole than by an individual.

One pathway to the order of the world leads through faith. In religion the human spirit directs itself to those creative powers that always commit us unconditionally, wherever we enter religion's sphere of influence. But one cannot speak about ultimate things. This is why all religion begins with the parable. It is the parable that, in a certain sense, establishes or creates the language appropriate for speaking about the complex realities of the world. The words of the parable are obscure: religion refuses from the outset to assign a scientifically precise, definitive meaning to words so that the meaning may emerge in each instance only to the degree that individuals in the course of their lives, and humankind in the course of millennia, learn to understand those complex realities of life. Sacred Scripture is capable of infinite interpretation; that is why it can outlast the millennia.

The common language created in the religious parables binds human beings together in a way more than any other common language. For people of the same tongue may indeed communicate well with one another about the activities and sufferings of daily life, while people of the same faith can understand one another in relation to the foundation of all that is and, therefore, as we know since Plato, also in relation to the order of values.

The ancient search for the meaning of life is answered in the parable which lies at the root of religion. On the one hand, the parable can speak directly of the meaning of life; on the other, the language formed in the parable may be of such a nature that the question of life's meaning is no longer posed in words.

A completely different pathway to the order of reality is taken by science or, more specifically, by empirical science. The hope expressed in science is that over the course of the centuries humanity may learn to speak about the whole of reality in a manner similar to the way the child learns ordinary language during the first years of life. While religion refuses from the outset to assign a precise, definite meaning to words—for that is how its basic formulae can outlast the millennia—science proceeds from the expectation that in the course of time words may eventually receive a precisely defined meaning. The language of science is changeable; it develops along with human experience and the basic disposition of science is skepticism. There are no final formulations in science in the same sense as there are in religion. Only when an overwhelming amount of experiential material has compelled us to devise very precise formulations and when, in addition, these have proven themselves again and again in multiple, diverse experiences, do we feel ourselves constrained to acknowledge that precisely these formulae represent the relevant domain of experience exactly. That is how they become a definitive component of scientific language. But if we ask where the limits of that domain of experience are and, concomitantly, the limits of the validity of the formulae that represent it, these questions can be answered once again only through new experiences.

History shows that humanity can find its way in reality in this manner too. Just as the child first becomes familiar with the simplest objects of daily life and then moves on to more complex concepts such as color, form, etc., finally learning to use abstract concepts as well, so humankind also first learned to order the most important areas of experience {of daily praxis}. That is how astronomy, geometry and statics and, simultaneously with them, always mathematics developed. Only after that did humankind push on to other areas that are more difficult to access. The beginning of this process is not certain knowledge but the practical success that is attained by tentatively stepping forward for the first time. Just as the child can learn words only in the constant interplay of acting, speaking and experiencing, so science develops in direct connection with practical application which, in the end, is the real measure for the correctness of the understanding gained. That is why physics and chemistry develop in connection with technology; geology and mineralogy in connection with mining; and biology, physiology and psychology in connection with medical science.

Thus, science's claim to truth is always derived from the object, for its language forms itself in correlation with the objective. Ideally the aim of a scientific representation is the "objective" representation of a specific state of affairs. It is assumed that the relevant condition can be sufficiently detached from us and from its representation that it in fact may be turned into a pure "object." But there are broad areas of reality that do not permit objectification in that sense, i.e., that cannot be detached from the viewpoint on which the research method is based on. But that does not mean that one therefore immediately loses the ability to represent them in the language of science. For even if something cannot be objectified in the manner set out above, that fact itself can be objectified and examined in its connection with other facts. Thus, in its striving for objectivity on a higher plane, the language of science can adapt itself to the other domains of reality as well. It is thus understandable, that the more this language turns to describing areas of reality that cannot readily be objectified, the more difficult it becomes. Children, too, learn first to name only their toys; only much later can they speak about joy or admiration or even about themselves.

Thus, the pathways taken by faith and science to the order of the world set out from precisely opposite poles. While science begins in the domain of reality where we seemingly may leave ourselves and our way of representing things out of the question, religion, by contrast, begins precisely in the domain the visible form of which must be created by ourselves alone, i.e., in the domain of creative powers where it is we ourselves who shape reality.

That is why the order of the world presented by religion has been juxtaposed as "subjective" to the order science presents, and is claimed to be "objective." It must be conceded that, historically speaking, a specific religion's claim to truth is limited to particular places and times, in contrast to such a claim by science. The gods of Greece forever ceased to rule the world ever since no more sacrifices were offered to them. By contrast, Archimedes' laws of levers [C18] are still valid today. But the ancient gods *really did* rule the Greek world in their time. Whoever wants to claim that this was so only in people's imagination may wish to infer by those words that in principle there may have been unbelievers then too, after all. But to put things that way would be to present a completely false picture of what people of that time really experienced. For example, whoever took part in the feasts of Dionysius could really be encountered by the deity.

The core domain from which we create reality ourselves constitutes for scientific language the infinitely remote singularity that even though it is indeed decisive for order in the finite sphere it can never be reached. Conversely, the language of faith cannot do justice to the domain of reality that is objectifiable and detached from us. For the words of that language have obtained their meaning precisely through their relation to us.

Religion alone can speak of the meaning of life. For "meaning" signifies that it is we who are addressed here—this is the point to which science cannot advance. That is why in science's language about the meaning of life [C19] one can only say with Bohr: "The meaning of life consists in that it is meaningless to assert that life has no meaning." Science offers so little comfort for that reason. But exactly that insight offers enough comfort to the wise person who has come to know that all ideas through which we seek to fathom life's meaning circle back to the point where they started.

The concepts "objective" and "subjective" designate two poles from which an order of reality can take its beginning. They also describe two sides of reality itself. Still, it would be a crude oversimplification to divide the world into an objective and a subjective reality. This merely black and white representation created much inflexibility in philosophy of the last few centuries. The evaluation of these two sides of reality also differed a great deal at different times. At times, one side would have only been regarded almost as a deceptive appearance. To our age it seems more natural not to raise the issue of evaluation at all here and, instead, to strive for a more refined and clearer classification of reality. Since that classification is to be scientific, it will proceed step by step from the objective to the subjective. The description and delimitation of the individual domains of reality is carried out with the greatest of care, as is appropriate to the natural science as it has developed over many centuries.

II

1. The Domain of Reality in Goethe's View

The order suggested by the development of natural science follows ancient patterns of thought that found ever new forms of expression at different times. We begin this reflection with a paragraph from the supplements to Goethe's <u>Theory of Colors</u> [C20]:

"All effects we become aware of through experience, in whatever form they are, are connected in the most coherent fashion; they flow one into the other and undulate from the first to the last like waves. That people separate them from and contrast them with one another and mix them together is unavoidable. But this had to give rise to an endless conflict in science. Rigid, divisive pedantry and blurry mysticism both bring about the same disastrous results. But those activities, from the most basic to the noblest, from that of the brick falling off the roof to the radiant insight of the spirit that dawns on you or that you share with others: they are linked together. We try to put that into words:

Accidental,
Mechanical,
Physical,
Chemical,
Organic,
Psychic,
Ethical,
Religious,
Ingenious."

The order of reality presented in this citation may serve as a model for the order modern science is searching for. But before such an order is implemented, it must be accurately determined what such a classification can and cannot mean.

First, this is clearly not a classification of objects (in the most general sense.) To be sure, initially, it may seem as if objects too could be ordered in terms of such categories. For example, rocks belong to the lowest domains: mechanics, physics and chemistry, whereas plants and animals are of the organic domain, while the highest domains are reserved for the human being, the one with a soul. But we have known for ages that "dead" matter also is incorporated into the organism in some kind of chemical transformations or other and can thereby take part in life. We know as well that the organism's functions are performed by the effect of laws of physics and chemistry and, finally, that psychic and physical processes are tightly coupled in their course. Surely that order is therefore not meant to represent a classification of objects.

Thus, if we take leave from the first idea and move one step forward, another view presents itself, that this order is about the ways substance behaves [C21]. One may conclude that one and the same substance can insert itself into the most diverse configurations. For example, the same drop of water flowing in a brook has to follow the laws of physics but, when mixed with the salts of the ground, is directed by chemical forces, and then when it is absorbed by the roots of a plant, it comes into the sphere of action that is the domain of organic laws, and so on. If one pursues this train of thought, it would appear that the above classification is about an order of configurations governed by laws of nature which, in a certain sense, are juxtaposed as guiding ideas to the "other," namely to substance. Goethe may have had something similar in mind, since he speaks of the different "activities." Yet, this view too must be corrected in one essential aspect. From the point of view of recent natural science, it is not possible in general to dissociate the concept of substance from that of the rules of nature. If one follows the development of the concept of matter in modern physics, matter, just like force, finally appears to be a kind of structure of space. That structure is subject to the laws of nature and, as a result of certain features of "invariance" in those laws, the word "matter" may be used in the description of processes. However, it is not matter but the law that remains constant as phenomena evolve.

Only when this step has been taken and it has been recognized that there is no "substance" that follows specific laws, but only complexes of connections which we can experience and that when we describe them we also occasionally use words like substance or matter—only then we may correctly understand the sentence that the sought-for classification is one that orders reality according to those connections.

By "domain of reality"—if the word domain is used in the special sense of classification—we mean a totality of complexes of configurations governed by laws. On the one hand, such a totality must present a solid unity, otherwise one could not justifiably speak of a "domain." On the other, that totality must be capable of being exactly delimited from totalities so that a classification of reality actually becomes possible. This raises the question how a totality of laws can become complete in itself and exactly delimited from laws of a different kind.

What must be kept in mind here first of all is that the language we use to speak about reality emerges in the interplay between acting and experiencing. To the degree that concepts get increasingly precise meanings in the course of time they also become conjoined with specific presuppositions about our ways of acting. This leads naturally to a certain closure of a system of concepts and of the domain of reality it signifies, since this system refers to a very definite way of behaving toward reality. That is why that system can no longer be used in cases where we decide to move into a different mode of action. On the other hand, the domain in which the system of concepts in question can be applied becomes problematic in any case since, after all, only through experience alone one can determine the extent to which the presupposed mode of action is possible at all. For example, concepts such as mass, location, velocity, straight line, plane, are used to describe the domain of reality that 19th century classical physics could comprehend. Thereby we have clear instructions on how to measure mass and location, how to construct or test a plane technologically. Only through those instructions do these concepts have meaning in physics. And only experience can decide where and to what extent these instructions can be carried out. Nevertheless, one may speak of a closed domain of reality that can be comprehended by means of these concepts.

Even though it appears fundamentally possible to distinguish clearly between different systems of concepts and domains of reality—given that they are associated with different ways of behaving and, consequently, different questions—little is actually gained by a general reference to such a possibility. For, on the one hand, it must be the aim of scientific observation not only to point out the domains of reality, but also to formulate with utter precision the totality of the complex of connections that signify each domain. On the other hand, the observation of the simplest processes of nature tells us that different domains always meet or intersect in those things that we make the object of our observation. Thus it should be possible to separate different domains by keeping in mind that different laws are related to the different ways of asking questions. It then follows that the different laws must also be closely connected to one another. For example, an animal's skeleton can be approached and examined from the perspective of physics, how as a static system it reacts to external demands, to pressure, push and pull. But it can also be studied as a member of an organism. But since it is the same object that sometimes appears as the sphere of action of physical laws, but at other times as that of biological laws, the two sets of laws must obviously be attuned with each other in such a way that no contradiction arises where they intersect. On the one hand, this mutual relation of different kinds of laws is an obvious thing because the laws are simply the expression of experience and experience begins with the recognition that there are things. On the other hand, this fact makes great demands on, let us say, mathematically formulated natural laws, by calling on two different formalisms to be so completely integrated that no contradiction can arise anywhere, even in their infinitely diverse consequences. In many cases, this can be accomplished only when one formalism is contained in the other as a restricted case. Thus, in the system of domains into which reality arranges itself for us—at least in certain parts of it—there is something like a ranking order in the sense that certain lower domains are contained in higher ones as special and simple restricted cases. That does not mean at all that the lower domain is not to be regarded as an autonomous totality of laws; other and simpler concepts may belong to the simple restricted case rather than to the original domain.

The order of reality we seek should arise from the objective to the subjective. That is to say, it is to begin from a part of reality that we can place completely outside ourselves, where we can completely set aside the methods that help us gain knowledge of its content. But at the apex of the order, there stand, as in Goethe's model, the creative forces with the help of which we ourselves transform and shape the world. The word "subjective" must not be misunderstood here in the sense that in the higher domains we are possibly dealing in part with matters that exist only in our feelings or in certain people, or that are a kind of imagined reality. In no way does the word "subjective" signify mere appearance. On the contrary, each of these domains consists to a high degree of features that can be completely objectified. "Subjective" is meant simply to suggest that, in a complete description of a domain's complex of c onnections, it may perhaps not be possible to ignore that we ourselves are interwoven in that web of connections. For example, when seeking completely to formulate the laws of atomic physics, we can no more ignore that our body and the instruments we use to make our observations are themselves subject to the laws of atomic physics; furthermore, in atomic physics, our *knowledge* of a state of affairs takes the place of a physical fact. How much more will the equivalent apply, let us say, in psychology. While the existence of the human soul is an objective fact that we can place completely outside ourselves, as an essential consideration, the fact that we are ourselves beings with souls still has to enter into the formulation of psychological laws. The forces of which the parables of religion speak are also "objective."

Thus, when it is said that the envisaged order of reality is to rise from the objective to the subjective, what this implies is that the process which provides information about reality, itself increasingly forms part of the relevant complex of connections that constitute the relevant domain. It may be that this might still seem too crude a statement if one already had as precise knowledge of the higher domains as is available to us of the lowest. But for now we {now} must content ourselves with knowing the general direction where the fundamental differences between the higher and the lower domains are to be sought.

One may object that the proposed division outlined in these preliminary remarks is not really about domains but about certain idealizations of reality. One may say, for example, that classical physics represents the idealization wherein we may completely leave aside the process of cognition that instructs us about reality. Thus, it looks as if taking this path would not lead to an order of reality but to an order of our understanding or cognition of reality. But the concept of reality indeed already presupposes not only the entity to be ordered, but also ourselves; it is, therefore, not surprising that when we are dealing with an "order," we cannot decide whether what we meet with is an order of reality or an order of our understanding of it. Furthermore, it is self-evident that every ordering arrangement of reality calls for an idealization, for reality encompasses us initially as a constantly flowing complex of connections from which we deduct certain processes, manifestations and laws only through the intervention of our thinking which, as such, idealizes. But, in the end, we must always bear in mind that the reality of which we can speak is never reality "per se" but a perceived reality even, in many cases, one we ourselves have shaped. It may be objected that this last statement concedes that there still is, after all, an objective world wholly independent of us and of our thinking, a world that runs or can run without our help, which we really envisage in our research. One must reply to this at first so plausible objection that the phrase "there is" itself already derives from human language, for which reason it cannot properly signify something that would not be related to our ability to comprehend. For us, "there is" simply only the world [C22] in which the phrase "there is" has meaning.

2. (Classical) Physics

The order of reality sought for here should thus start from that "most objective" domain of reality we can speak of without referring directly to the methods with which we learn of it. That domain contains primarily that part of the world that permits its description in scientific form in terms of so-called "classical physics". The first completely developed discipline of this classical physics was Newtonian mechanics. It became the model for the way a totality of connections through laws of nature could be formulated through mathematics. As such it became extremely significant for the development of natural science. For that reason, classical mechanics will be the first area dealt with here.

(a) Mechanics

Newtonian physics begins with the concept of substance [C23]. It presupposes that every body consists of a specific amount of "matter" that is not subject to form or changes in motion. This substance can be called "mass," *quantitas materiae* in Newton's words. This mass pertains entirely to the body as such; it is completely independent of the methods by which it can be determined. Obviously, this first assumption represents an idealization that is not realized precisely in nature, since the bodies accessible to our observation are in a constant process of reciprocal interaction with other bodies which can continually cause small changes in the mass. However, the fact that those changes are minimal still allows the concept of "mass" to be introduced and subsequently refined as the laws of mechanics are formulated, so that these small "changes of the mass" can also be taken into account.

Mechanics presupposes further that every body (or its individual parts) is assigned a specific position. Space and time are seen as two firm, mutually independent patterns of order into which the world's processes can be arranged. Newtonian theory renounces from the outset the idea, already proposed in the days of Greek philosophy, that space and matter might be connected with each other. For example, in this sense, one might see space as upheld by matter's structure, or hold that matter is to be considered as a structure of space. Newtonian theory was instead content to objectify space and time in the form we know simply because experience teaches us that such an objectification is possible in a wide range of experiences. Here it is presupposed that Euclid's geometry is "valid" in space. On the one hand, what we have here is a supposition that cannot be tested by experience. For when, let us say, a measurement taken with physical instruments shows a deviation from the laws of Euclidian geometry, one can always attribute this to the physical properties of the instruments. Seeking to determine whether metal sheets are perfectly plane, a mechanic in the workshop places two of three sheets on top of each other and, moving one over the other, watches whether they always touch everywhere. Repeating this procedure with every sheet, if the procedure is successful, he establishes the validity of Euclidian geometry in the planes of those sheets. On the other hand, this supposition of Euclidian geometry is the outcome of experience, inasmuch as it becomes obvious to us because we have so much success with that supposition in our experience. The mechanic *can* in fact so burnish the three metal sheets that, when stacked, they touch everywhere no matter how they are moved; this is true at least in relation to the precision his tools can offer. It is our *experience* that this is so and nothing gives us the right to believe that the described process can also be carried out with ever increasing accuracy.

Newtonian mechanics also presupposes that there is a firm scale of time independent from position. Here too, then, time is thought to "move" objectively independent of all events and it is assumed that this time can be observed by measuring it with randomly located, coordinated clocks. This presupposition, too, proves itself valid in a wide range of experiences.

Thus, Newtonian mechanics looks at the processes of nature insofar as (-independently of all methods of observation-) one can attribute to bodies solid mass and specific location, ascertainable at every moment in time, within a space of Euclidian metrics. Experience then determines the extent to which these concepts of mass, location, velocity, etc., suffice in the representation of nature.

In mechanics, material "mass," (as that which "moves,") is juxtaposed as that which "moves" to immaterial "force." This is a curious contrast insofar as force, like matter, is said to be an objective effect fixed in space and time. It is thus difficult to understand why force is understood as something quite different from matter. Despite the successful results achieved with the idea of such an immaterial force acting upon a body, there arises the need to let like act only upon like, that is, to regard force itself again as a form of matter. Indeed, this step was taken in the nineteenth century in the development of the theory of electricity. More will be said about it later.

The special preferential place that mechanics has in relation to all other parts of a scientific representation of nature is based on two factors. The first is that the object of mechanics is the objective behavior of material bodies in space and time, in other words, something especially simple and tangible. The second factor is that no condition of matter can be conceived in which the action of mechanical laws does not become visibly manifest. Even though there are indeed conditions of matter in which the action of neither chemical nor biological laws is directly manifest—if one thinks of the hot gases in the interior of stars, for example—the operation of mechanical laws remains visible as long as one can speak of matter at all. Thus, mechanics can be described, on the one hand, as a particularly limited form of representing nature because it studies nature's processes only insofar as they can be completely objectified. On the other hand, mechanics can be regarded as the most encompassing manifestation of the governance of the laws of nature, appearing everywhere matter is encountered at all.

The laws of Newtonian mechanics are depicted comprehensively by a mathematical formalism. In this formalism, an infinite number of diverse conclusions may be drawn from fundamental presuppositions, all of which follow necessarily from these presuppositions. Hence, one might say that the outcome of any one mechanical experiment can be predicted with mathematical certainty. One may begin by objecting that the presupposition concerning the validity of the laws articulated in mathematical forms is, after all, a matter of experience. From the fact that the sun has risen every day thus far one can at no time logically conclude that it must rise again the next day.

But, practically speaking, this objection only means that we have to reduce the degree of certainty in what we expect the outcome of the experiment to be. The result of the experiment can be presupposed with the same degree of certainty we have in knowing that the sun will rise the next day. This degree of certainty always suffices for us. Of course, the question whether we are talking about a "mechanical" experiment at all results in another reduction in that certainty. Of course this is always the case only to a certain degree of approximation. Generally speaking, one may judge from the way the experiment is set up with which degree of certainty one can express the experimental presuppositions in the concepts of mass, location, velocity, etc. The results of an experiment can be predicted from the laws of mechanics with the degree of certainty that corresponds to degrees of precision. The following example may help to elucidate these different gradations of certainty. An inventor, striving to construct a perpetuum mobile [C24]—if he does not limit himself to mechanical processes or to those where we know completely what laws apply to them—can be told only that, according to the laws of nature we know thus far, it is probably impossible to construct such a perpetuum mobile. If the inventor tries to construct it strictly by mechanics, one can declare the effort senseless with the same degree of certainty we have in knowing that the sun will rise tomorrow. But if the inventor undertakes to construct the perpetuum mobile in his mind, using Newtonian laws, he commits a mathematical error.

(b) Electricity and Magnetism

The development of the theory of electricity was responsible for the first decisive step forward in classical physics beyond Newtonian mechanics. The particular forces that are present between electrically charged or magnetic bodies are the object of this discipline. At the same time, it enables a quite general analysis of the concept of force which was still somewhat alien to Newtonian physics. In the theory of electricity, force is objectified and fixed in space and time by the concept of the "force field" in a manner similar to the way matter is objectified and fixed in mechanics. Force appears not only as an effect of one body on another, but is itself a process in space and time that can detach itself completely from all matter. Understanding the autonomy of force helps clarify the intrinsic relationship between force and matter which were eventually most clearly articulated in the natural laws that were discovered only at the beginning of the twentieth century in connection with the special theory of relativity. From the perspective of what we know today, it does not appear anomalous, for example, to speak of radiation, that is, the electromagnetic force field, as a special form of matter. Matter can change into radiation and radiation into matter. The principle about the conservation of matter is expanded into one about the conservation of energy, and energy can present itself in the most diverse forms: as radiation, motion, weight.

This development from mechanics to the theory of electricity met initially with resistance from many natural scientists because the idea of an autonomous force field apparently leads to the dissolution of the primitive concept of substance in Newtonian physics. Thus, there was no lack of experiments to reinstate the original concept of substance by introducing a hypothetical ether [C25] which was to be perceived as a carrier of the electromagnetic fields that appeared to be autonomous in space. But a substance that cannot be localized in space, that we can neither feel, weigh nor see, like ether, deserves this name even less than the electromagnetic force field. As a result, researchers had to decide to drop the primitive concept of matter. After all, the language we use to speak about nature is created in the interplay between action and experience and is not dependent on our often historically conditioned desires.

Thus, even though our experience with electricity and magnetism forced science to drop a fundamental concept that had for long been seen to be the surest pillar of the house of natural science, nature itself helps us in fullest measure in our need to objectify and project all that happens into an objective process in space and time. The theory of electricity can be seen in a sense as the completion of this program. For since it was incorporated into the system of physical laws, all we are now dealing with is a unified, objectively transpiring occurrence in space and time that encompasses all processes of nature independent of all observation. Depending on the particular situation and on questions we pose, we can associate the occurrence in a quite specific manner with the concepts of matter, energy, radiation, force, etc. The entirety of the laws governing this occurrence can be summed up within the designation of "classical physics."

The processes occurring in space are completely determined here in their duration in time by the condition prevailing at the start of the process. Thus, classical physics perfectly satisfies our desire to have the occurrence ordered in terms of cause and effect, that is to say, to perceive all that happens as causally conditioned. Classical physics idealizes processes in the world by regarding events as occurring in isolation in space. Because it is believed that they can be ruled out in principle, no consideration is paid in classical physics to the mutual connections that must exist between the processes in question and the surrounding world that allow us to study the processes and have them become part of our world. If it is possible to uphold such isolation, then we can also gain perspective on all causes. In the part of the world thus isolated, the insistence on causality can be satisfied only by the idea that everything is completely determined.

Furthermore, something that used to be said about mechanics applies to this system of classical physics as a whole. If the concept of matter is broadened in the way suggested by the development of the theory of electricity, then no condition of matter can be imagined where the action of the laws of classical physics is not clearly manifest. However, such conditions do exist where the complex of connections from the chemical or biological domain is not immediately detected.

The entirety of connections governed by natural laws that we call classical physics is "closed" in itself. That is to say, classical physics can be articulated in a system of concepts and axioms without contradictions and that can be called "complete" in a certain sense. It is complete, to be sure, insofar as it is possible to add new concepts to the system, but then those new concepts are in their formal connection to the system necessarily so closely related to already existing concepts that they function more as a new "representation" (in the mathematical sense) of the old concepts. It is in this sense that we speak of complexes of connections in their entirety as "closed." The laws of classical physics automatically get separated from other forms of connections by the demand that we should always be dealing with the objective changes in time of certain quantities in space. If the dimensions of "classical physics" are set this widely—naturally, there is no historical but only a conceptual justification for this—then its domain also includes various complexes of connections that were discovered only very recently and that share the feature of fixing objective processes in space-time with the earlier components of physics. The discovery of matter-waves by de Broglie [C26] may serve as an example; they may be understood as objective wave processes in space-time if one ignores the corpuscular nature of matter. If one does not ignore the existence of elemental particles, then the patterns of the laws relevant here cease to be part of the domain of classical physics but belong to what falls under the comprehensive concept of quantum theory. This example demonstrates that it is not *things* but the *complexes of connections* that can be ordered according to different domains.

(c) The Infinite

Among the elements of classical physics are its very definite presuppositions about the structure of space and time. The most important aspects of schemes of order, thought to be independent of all matter, are the following:

1. There is no fixed point in empty space; rather space has the same structure from whatever point it is viewed.
2. The same applies *mutatis mutandis* to time.
3. There is no fixed direction in empty space; rather space has the same structure from whatever *position* it is viewed. In mathematical terms, space is isotropic.
4. In empty space there is no uniform straight movement that is preferred over any other such movement.

Because space is three-dimensional, these four statements contain a total of ten basic properties of space and time, if one counts each statement separately for every dimension of space. These ten basic properties form the actual foundation of classical physics. The entire structure of the mathematical system in terms of which classical physics is formulated is conclusively determined by those ten "invariance properties." We may say that those properties are merely a more precise expression of what we understand by "empty" space. For if, let us say, certain points were fixed in space, then these points could, from the perspective of field-theory's concept of matter, be interpreted logically as positions of matter.

Two further significant presuppositions are added in classical physics to those ten basic properties: space and time are independent and Euclid's geometry is valid. It is only the totality of these presuppositions that completely determines in what forms of space and time we view the processes of the world.

The very fact of the existence of such forms of perception raises the question of how the world looks when we think of spaces that are either enormously large or very small in comparison to those where our experience takes place, or when we think of a past time countless years ago or of a likewise distant future. The question of "the end of the world" is likely as old as human thought; different ages have answered it very differently.

In earlier centuries, when such problems were still surrounded by myth, humans thought of the world as a flat disc. The ocean swirled around the inhabitable lands and were for all practical purposes "the end of the world."

This ocean represented infinity in that everywhere it stretched in equal and empty measure into endlessness, a monotonously uniform desert where no one could imagine how it could end, since no land on the other shore was there to offer hospitality to the sailor. Above the ocean the sky formed a vault. Later, when the spherical form of the earth had become known, the sky was thought to be structured by the crystalline spheres of the planets and eventually fenced off by the final sphere, that of the fixed stars, from actual infinity, as it were, the eternal fire, that can even fill the entire space beyond the world uniformly and without limit. Thus, people then were satisfied that the world accessible to our senses is limited by certain boundaries against an endless world that begins beyond them. And even though this notion of an infinity beginning on the other side had to arise from the manner in which things were perceived, it was most likely linked to the idea that these areas are no longer within the competence of our ordinary perception. Space in the ordinary sense was related to our earth and whatever space there was beyond the boundaries was of a different kind and seemed by its infiniteness alone to enter into an unmediated connection with the creative powers that we mean when we speak of religion.

Similarly, the span of time to which the word time may be applied in its ordinary sense, was thought of as limited by the creation of the world and its eventual demise. And there was a definite feeling that that was where questions must stop; one must not inquire about what was there before the world was created or what will happen after its demise.

This mythical world view, long since replaced in numerous details by one closer to the current stage of knowledge, only disappeared completely from human consciousness after the time of the first circumnavigations of the world. Since then, we know that one cannot travel, at least here on earth, in any case, to the end of the world because every road we set out on and continue traveling further in a certain direction eventually leads back to the starting point. But about the same time, it was also recognized that where earlier the sphere of the fixed stars was thought to be was only the beginning of spaces of utterly unimaginable size opening which our eyes can penetrate as we look at the most remote stars. The idea of a limit to the world thus gradually disappeared from the scientific world view and those remotest spaces were judged according to the forms of perception we are accustomed to in our understanding of the world, that is, according to the rules of Euclidian geometry. Thus, the idea arose that by proceeding in a straight line in a particular direction, we would enter into ever more distant spaces and beyond them into the infinite. In so doing, it would be the task of research to determine whether we would then eventually perhaps encounter utterly empty spaces or ever new systems of stars.

In the course of the development of natural science, the thought that the laws of nature and of our forms of perception may be applied to the infinite was eventually also transferred to time. In this case, too, the notion that the world endures only from the day of its creation until its demise disappeared and was replaced by presuppositions that pointed to something like an eternal periodic change of all events, that is, to a stationary condition of the universe seen in very large terms.

The complete independence of the space-time structure from matter, upheld as the foundation of classical physics, necessarily gave rise to the idea that, in principle, the same conditions would be encountered in the minutest spaces and times as in the macrocosm. This led logically to world views in which, for example, the smallest particles of matter were each interpreted to a new cosmos of so much smaller a world and, similarly, our visible cosmos was seen as a minute piece of matter of an even much larger cosmos—(and so forth into the infinite.)

The assumption that there are indivisible, ultimate building blocks of matter, in other words, the hypothesis of atoms, appeared like a foreign body in classical physics. There, in the best of cases, atoms could play a similar role in the construction of matter as, for example, cells do for the construction of an organism, although perhaps it could be conceded that perhaps atoms, too, are practically indivisible. But basically, like every other piece of matter, the atom had to be divisible, that is, it was supposed there could be no atoms in the actual sense.

The fundamentals of classical physics and its idea of the independence of the space-time structure and matter lead to such a view of the world. As has already been repeatedly emphasized, the hypothesis of independence is clearly an idealization, a formation of concepts that holds up so well when compared with daily experience that it automatically becomes part of the language with which we speak about nature. Yet, this hypothesis can hardly be supported by the experiences of very large or very small spaces and time-periods. On the contrary, the very existence of atoms—which in the course of scientific development became ever more apparent in chemical experiences—had to be understood as a counter-argument to the independence of space-time structure and matter. But, curiously, the revision of this classical view of the world was initiated by a much more profound feature of reality, accessible only by means of very subtle observations.

It has been observed that the velocity of the transmission of light in empty space is completely independent of the color and of the state of motion of the source of light. Only when one assumes that the velocity of light has at all times a constant value relative to an observer, independent of his movement, can that experience be brought into agreement with the principle that there is no uniform straight movement in empty space that is distinct from any other similar condition. Experience validated this conclusion. But this result is totally incomprehensible on the basis of Newtonian physics, where space and time are held to be completely independent, and thus it holds that the velocity of a ray of light relative to the observer must necessarily depend on the speed of this observer. As a result of precise experimental and theoretical analysis of these facts finally led to abandoning the thesis of the independence of space and time. Another concept took its place: We call events "past" when we can come to know something—at least in principle—about them; we call processes "future" when we can still influence them (again in this sense of "in principle"). In Newtonian physics, these two groups of events—past and future—were separated by an infinitely narrow gap of time that we call the present or, more precisely, the present moment. But, according to the results of the "special theory of relativity," those groups are separated by a gap of *finite* dimension where the duration of the "present" or the "synchronous" field of events becomes longer the greater the distance is between us and where those events take place. Einstein was the first person courageous enough [C27] to articulate the dependence of space and time and he articulated it in mathematical terms. Later experiments confirmed over and over again the conclusions of this new view of the space-time structure with greatest precision so that its accuracy can hardly be doubted any more.

With that discovery came the first breach in the classical space-time conception. It enriched the science of physics by the recognition that even the simplest fundamentals of physics, the forms of perception, cannot necessarily be used for processes that are far removed from daily experience, but, rather, that they must continually be re-tested by experiments in these new domains.

Scholarly research had been reminded in an unmistakable manner that, as the edifice of scholarship grows higher and higher, its foundations also have to be laid more and more deeply if they are to support the weight of this edifice. For the ground on which this edifice rests is, after all, not the rock of a sure knowledge that precedes all scholarly understanding but, rather, it is the fruitful soil of the language that is formed out of action and experience.

Reality and Its Order 49

The special theory of relativity replaces Newtonian theory's view of space and time with a different view, but one that is no less definite. It allows for just as much independence of the space-time structure, on the one hand, and matter, on the other, although the inclusion of the speed of light already signals a close relation between that structure and the physical processes. Only when physicists attempted to incorporate gravitation into their system of knowledge did they give up their belief in the independence of space and time and come to assume that the geometry of the world is determined by the distribution of matter in very large spaces and times. This "general theory of relativity" is probably [C28] not yet proven by experiments as reliably as the "special theory" since there are only few relevant experiments that can be conducted with sufficient accuracy to be regarded as a touchstone for the theory. But if one wants to attempt to sketch a picture of far-distant spaces and times at all, this only makes sense on the basis of a theory that at least does justice to all previous experiences of space and time, which is to say, for the time being on the basis of the general theory of relativity, for it links our experiences of electromagnetic waves with those of the phenomena of gravitation without contradiction.

If one proceeds from the concept of space and time presented by this theory, one has to assume that the geometrical conditions in remote spaces and times depend on the distribution and behavior of matter in the macrocosm. And if, in addition, one accepts what astronomers have learned in the last decades about the distribution of stellar matter in space, one comes close, oddly enough, to a world view that seems akin to some features of the mythical world view of the pre-scientific age. For the experiences of astronomers suggest that the world's space is finite. This is not meant, of course, to be true in the simple sense of that earlier view of the world; we will not arrive eventually at a boundary somewhere far away from our earth but, as with travels on our globe, as we proceed in a straight line into ever greater distances, we should eventually return to our starting point. The road traveled that way is said to be finite and can provide a basis for determining the size of the universe. From what we know thus far, one can conclude that it may be a journey of several billion light years. Other observations in astronomy lead to the conclusion that the condition of the universe some five billion years ago must have been very different from now; the world's matter then was apparently compressed in a much narrower space under extremely high temperatures. Only much later did stars and stellar constellations come into existence. This perception has occasionally been expressed as follows: The universe has only existed for five billion years [C29]. But such an expression is always intended to include the proviso: As long as the concept "year" can be applied to the early stages of that process of development. Thus, what we are dealing with here is a statement in scientific terminology that should be read something like the statement: The absolute zero, the lowest possible temperature [C30], is around minus 273 °C. Even though such a statement accurately describes certain experiences, it would be just as possible to devise a temperature scale that reaches down to negative infinity degrees, (that is, more accurately, it cannot be reached at all,) without thereby having to change the understanding of temperature we have accessible to us in our daily experience. Similarly, the time unit used for measuring time in an earlier stage of the universe becomes problematic when the advance of time could not yet be measured by the regular circular course of the celestial bodies.

Thus, the experiences of astronomy indicate that the structure of space and time manifest to us is, in some degree, valid only for spaces that are very small in comparison to distances of several billion light years and for times that are very short in comparison to several billion light years. It is true that we do not encounter boundaries in spaces still farther away and in still earlier times, but we do encounter a changed structure of space and time which is equivalent to a finite expansion of the world in space and time.

These inconceivably great distances modern astronomy talks of in this view of the world are seen in a yet different light, incidentally, when considered from the perspective of the mutual dependence of space and time. For example, one may ask how long it would take a space traveler moving in space at bearable levels of acceleration—and such an idea is perhaps no longer utterly absurd, given today's technology—to travel from our stars to systems of stars huge distances away. According to a theoretical estimate, given the peculiar space-time structure described mathematically in the special theory of relativity, a period of a few decades would suffice to cover the distance to the most distant spiral nebulae known to us. It would only be for the observer who follows that space voyage from the earth that this amount of time will seem unimaginably long.

When we turn again from the very large spaces to the smallest ones, a view of the world that regards the atom itself [C31] as a small cosmos will no longer appear credible to us. Rather, modern research thinks it entirely plausible that other laws, and another structure of space and time as well, are valid in very small spaces and very short times than those in the sphere of daily experience. Today's natural science has come to terms with the idea that quite definite expansions of space and time periods in the world are marked as distinct from others. Such distinctions would be incomprehensible if what is being dealt with here were only certain scales of things. For why should the same things not be able to exist in any number of smaller or larger sizes? The distinction becomes comprehensible, however, when those fundamental units of measurement appear in *the laws of nature*. For a simultaneous existence of two such laws with different values of the constants would be a contradiction.

In the development of physics from Newtonian mechanics via the theory of electricity to the theory of relativity and the view of the world of modern astronomy, an uncritical application of certain concepts was sacrificed as one strove for unification and simplicity. Those concepts had become components of our language of natural science on account of their fundamental significance. Since this development occurred, the words "matter," "force," "structure of space and time" seem to represent only different sides of the same event. Thus it follows from this unification of concepts that none of those concepts can be used without reservations in their simple, original meaning, unless we are dealing with processes in the sphere of daily experience.

Although the pursuit of knowledge has, as a result of this development, moved away in various ways from the idealization of the processes of nature that is at the core of Newtonian mechanics, it has nevertheless continued to this day to hold fast to the "classical" idealization in one respect: It studied processes in nature only insofar as they can be objectified, that is, projected away from us as objective events in space and time. It is this fact that gives us the right to declare the part of reality discussed here under the disciplines of physics an ultimately unified complex of connections. What the development of natural science makes abundantly clear to us is that such a unified domain also manifests internal structures and that through more in-depth idealization other, simpler sub-domains may be separated out from it.

3. Chemistry

In the language of classical physics that we use to comprehend the things of this world, words as simple as matter, force, movement, etc., seem very poor in comparison to the infinite fullness of phenomena. The next expansion of language that emerges here as a matter of course has to do with the sensory properties of things. We describe material objects as warm or cold, solid or fluid, of this or that color, tough or brittle, hard or soft, sour or alkaline, salty, combustible, etc. Human language has at its disposal a great quantity of words to denote properties of things which do not fit into the scheme of concepts consisting only of terms like movement, force, matter.

These properties can be objectified just the same as the processes of classical physics: A body as such is warm irrespective of how this might be established. All the properties named pertain to that body objectively, regardless, for example, of the method of observation. So, as long as research is content with accepting such properties into scientific language, it expands only the objective domain we have called classical physics. Or one might well say that another domain of properties appears besides the domain of classical physics. Both domains have in common that we are able to speak about objective things and processes in space and time. However, aside from its mechanical properties, a thing possesses others as well, such as a color, chemical behavior, etc.

We have learned from many different experiences that there is still a close connection between the mechanical behavior of objects and their properties. Already the ancient theory of atoms was based on the general idea that the sensual properties of substances could be explained in terms of position and movement, that is, in terms of the mechanical behavior of the atoms. At the beginning of the 19th century, nascent scientific chemistry and the development of theory of heat forced researchers to examine more closely the relation between the mechanical behavior of the smallest particles of matter and the properties of bodies. The first area where those relations were fundamentally clarified was theory of heat.

(a) Heat

Observing very small particles of matter microscopically shows that when small particles are moving freely, let us say in a fluid, they are constantly in an irregular flutter motion. This movement increases when the fluid in which the particles are suspended is heated. Since it is apparent, furthermore that this motion is dependent only on the particles' mass and size and that it becomes stronger the smaller the particles, it was natural to generalize the observation and assume that as heat increases there is always a correspondingly increasing motion of all particles, that is, ultimately of all atoms of a body. One could indeed conclude that heat is "nothing more than the irregular state of atoms in motion." [C32]. This formulation was not meant to remove heat as a sensation from scientific language. However, the concept "heat" was to be divided into an objective and a subjective component. Insofar as heat was understood as an objective process in space and time, it was to be presented as atoms in motion. In addition to this, it can also be perceived as the content of a sensation and as such as an object of observation. But then it is a part of human experience and no longer an objective process in space and time. During the 19th century physics brought the plan to completion that was envisioned in the assumption referred to; only in one aspect [C34] did experience force that plan to be rather decisively altered.

The studies of the behavior of heated bodies led initially to the development of the so-called phenomenological theory of heat which in turn, using the concepts of quantity of heat and temperature (together with the concepts of classical mechanics), established an order of phenomena that was appropriate to classical physics. The attempt to link heat and the motion of atoms also showed that quantity of heat may be understood simply as the content of the energy of the atoms' motion. Thus, an entity known from mechanics, namely energy, takes the place of "quantity of heat."

But there is no simple mechanical substitute for the concept of temperature. Rather, it turned out that temperature does not signify a system's mechanical property but, instead, the degree of our knowledge of the system. Knowing a body's temperature means we know that the body is a randomly chosen sample, as it were, from a large totality of similar bodies, with "temperature" saying something about its distribution within the totality (the so-called "canonical totality." [C33]). Therefore, temperature is not something belonging to the mechanical system as such but denotes our *knowledge* of the system. Here atomic theory of heat differs fundamentally from classical physics. For the first time, our knowledge of a system has become a factor of physics. As a result, we have to place atomic theory of heat into the next domain of reality where processes can no longer be unconditionally projected into space and time as objective events. Another immediate consequence of this circumstance is that it is impossible within atomic theory of heat unambiguously to determine future processes. For when a body's temperature has been established, all it means is that we have a certain degree of knowledge and of ignorance concerning the mechanical behavior of atoms about whose future motion we can at best indicate that a certain process will probably take place.

Also, the experience of 19th century research into heat leaves us with the following, peculiar situation: On the one hand, one can regard heat and temperature as new, objective qualities of matter and study their laws. But then, one disregards the relation between heat and the motion of atoms, something that is directly and clearly discernible from a host of facts. Or one can idealize the motions of atoms in the sense of classical physics. Here, too, it would seem that we are dealing simply with objective processes in space and time. Finally, one may conclude from many experiences that the concepts of "heat" and "temperature" must mean something in relation to the mechanical motion of atoms. Then one may indeed objectify the concept of quantity of heat, but temperature states something about the *probable* condition of the atomic system, that is to say, about our *knowledge* of it. Thus, temperature no longer pertains to bodies as such, which reminds us anew that the world we can speak about is not the world "as such" but the world as we know it.

As described, the situation in theory of heat could possibly lead one to the following supposition, namely that concepts like "temperature," formed to describe properties, should apply only in a limited sense—for example, only to systems of innumerable atoms, under suitable conditions—but that they would not really describe the real behavior of bodies. They might be thought to be much more statistical concepts (like "the age of humans") that could be used in accordance with their original determination only under suitable conditions. Behind the world describable in statistical concepts would presumably lie an objective reality, namely the position and motion of atoms. Mechanical concepts would, therefore, have pre-eminence over properties since they were thought to describe the actual events of the real world.

But that supposition has proven to be incorrect in light of what has been discovered over the past decades about the connections between chemical properties and atomic motion. Before addressing those discoveries, we must stress, by the way, how unsatisfying it would be to think that amidst the great wealth of human language the concepts of mechanics alone would be appropriate to describe the "actual" behavior of the world.

(b) The Laws of Chemistry

Even at a surface level, the development of chemistry has some similarity to that of theory of heat. Both developments proceeded in tandem during the final decades of the 19th century. Chemistry also begins with a phenomenological description of the way things are connected, that is, with an objectification of observed chemical properties, from which emerge concepts such as sour-alkaline, compound-element, solid-fluid, crystalline-amorphous, etc. The hypothesis of the atom proved itself to be the most natural method for ordering the discovered connections. The atom has to be ascribed specific forces, the so-called valences, by means of which it can attract neighboring atoms. Together, the concepts of atom and valence provide the primary framework with the help of which the house of chemistry can be erected.

Experiences in electro-chemistry gave rise to the assumption that atoms are capable of taking on certain electrical charges. The most natural geometrical extrapolation from this situation was once again the hypothesis that atoms are structured from electrically charged elementary building blocks. The development of chemistry thus automatically led research to examine the relations between chemical properties and the mechanical and electrical behavior of elementary particles. It was not absolutely necessary for the advance of chemistry to take the hypothesis of the atom literally in this way. Since the size of atoms does not play a role in most of the laws of chemistry, the concept of the atom could be viewed strictly as a working hypothesis. The chemical forces were accepted as such and not explained further. But increasing refinement of the tools of observation brought atoms and elementary particles directly into the accessible field of experience so that it became impossible to avoid the question any longer of the connection between the laws of chemistry and the mechanical behavior of elementary particles.

This situation became the soil on which Bohr's theory of the atom grew. It comprises the result of all the experiences that inform us that the atom behaves in many respects like a mechanical system of electrical elementary charges. The atom of a chemical element cannot be regarded as the ultimate, indivisible unit of matter, but has to be seen as a composite of electrical elementary particles: Electron, proton, etc. (It is perhaps an unfortunate coincidence that in the course of its usage over time, the word atom has become associated with the smallest unit of matter of a chemical element, whereas, according to the Greek meaning of that word, it should be reserved for elementary particles.) In conjunction with modern quantum theory, Bohr's theory solves the problem [C35] of tracing the chemical properties of matter back to the mechanical or electrical behavior of the atoms. But this observation is valid only on condition that the mechanical behavior of the atom's elementary particles is not dealt with in terms of the conceptual means of classical physics. It is perhaps more accurate to say that Bohr's theory traces the chemical behavior of substances back to simple rules that apply to elementary particles, rules that can be established with mathematical exactness like the rules of classical physics to which, in addition, they have a close, connection in content. The manner of that "close connection in content" is decisive for the relation of this second domain or reality of chemistry to that of classical physics.

The laws of quantum theory may be presented as follows: The "state" of an atomic system is describable in terms of certain "quantities of state" or "functions of state." But those quantities do not directly represent a process or situation in space and time, like those of classical mechanics; they are not simply the locations and velocities of the particles that characterize a state. Rather, they have a certain relationship to the concept of temperature insofar as they generally provide us only with information regarding the *probability* with which we might anticipate certain locations and velocities of elementary particles if we undertake to observe them. Moreover, these quantities of state are more polygonal than those of classical physics. Atoms do have properties different from those of the mechanical systems of classical physics, especially the ones that have to do with the "sensory properties" of things. The quantities of state also contain information concerning the probability of certain quantities pertaining to those different properties, for example, the probability that they are of a specific color or that an atom has such and such chemical affinity. The quantity of state can itself not be tied to a tangible concept like location, velocity, color, temperature. It can be analyzed only with regard to tangible properties such as these and it then designates the probability that the property under observation will have quite specific values. It may well be, then, that in special cases the degree of probability will approach certainty so that one may say that the quantity of state represents a specific objective property of the system. But even then the quantity of state represents *more* than the description of the said property and this "more" is not an "objective" fact.

Two features of this situation are particularly important. One is that the quantity of state and everything it expresses does not in itself represent an objective fact in space and time. The other is that it is necessary through observation to analyze the quantity of state that establishes its connection to reality.

In relation to the first point, the impossibility to objectify these quantities of state in the ordinary sense is due most clearly to the abstract, mathematical character of those quantities. They are frequently represented formally in terms of functions in multidimensional abstract spaces that, as a result, cannot directly signify a process in space and time as we perceive them. The concept of state in quantum theory is far more abstract than, for example, the concept of temperature in heat theory which is akin to a sensory perception. But it is this abstractness that yields the richness that allows the quantity of state to make connection with sensory properties of very different kinds and to say something about them. Thus, the quantity of state in general denotes not a certain property of the system accessible to the senses but a specifically defined range of possibilities [C36] pertaining to all properties accessible to the senses.

It is only through the act of observation that something *objective* is created from these possibilities. It follows that this act of observation and the intervention necessitated by observation are a decisive feature of quantum theory and its subject matter. Observation generally alters the state of the system. It does so, on the one hand, through the very intervention that makes the observation possible. In addition, in the area dealing with the discontinuous changes of the smallest units of matter, this intervention can no longer be minimized at will nor its repercussions accurately determined. On the other hand, every observation similarly alters our knowledge of the system. As the content that can rationally be associated with the concept "quantity of system" is, after all, knowledge of possible or probable behavior, observation discontinuously alters what we must call "state."

The intervention made by the act of observation causes, furthermore, that not all of the system's properties can be simultaneously objectified. Rather, the individual properties are frequently in "complementary" relationships. What this means is that objectifications, that is, the observation of a specific property, excludes knowledge of certain other properties. The observation of a specific property of the system so alters the state that something learned in earlier observations about the value, or the probable value, of another property is lost in the process.

We can now more accurately describe what we can know about an atomic system, that is to say, about the content of the function of state: The function of state connotes, in the first instance, only probabilities that a certain result will be found when a property of the system is investigated. For individual properties probability may equal certainty. In this respect, the quantity of the state then denotes an objective fact. But there are always other complementary properties of the system about which knowledge of the function of state yields no objective facts. For those other properties the function of state once again indicates only the probability that something specific may be found when an observation of them is undertaken. But such observation is made possible only by an external intervention. It *alters* the system in such a way that the *new* property being looked for can be objectified, but at the price of losing the objectification of the previously known property. Thus, the result of the new observation leads to new knowledge and, being presented as such, to a corresponding new function of state, that in turn objectifies properties of the system other than those of the earlier observation.

This state of affairs can be made more comprehensible by means of examples. Let us say that the chemical behavior of a substance is known. For example, given that a certain gas is chemically pure hydrogen at a known temperature; we thus know that when combined, say, with oxygen and an release of a very specific amount of heat, hydrogen can combust to water.

These assertions convey a certain degree of knowledge about the state of the molecules that can be recorded in a function of state. This state of the molecules does not permit the objectification of the geometrical properties of the atom, that is, of the location or motion of the electrons that comprise the molecule. We may say that the movement of the electrons in this state is unknown in principle. But it is probably more accurate to say there exist no such motions in this state since we would take "motion" to mean an objective process in space and time. Of course, one can conduct experiments that would inform us about the electrons' motion in the molecule. But such an experiment so alters the state of the relevant molecule that it is possible henceforth to speak of positions of electrons in space and time, but the chemical behavior of the molecule can no longer be objectified. For example, the heat of hydrogen's fusion with oxygen is now unknown in principle; the result of an experiment seeking to determine it would be predictable only statistically. One may put the matter radically by saying: It makes no sense in this new state to speak of a specific heat of fusion.

It used to be said that Bohr's theory attributed the chemical behavior of matter to the motion of the smallest particles, the electrons. Now we understand that the state of affairs depicted in quantum theory may be expressed in a seemingly opposite way. It is fair to say that quantum theory has actually demonstrated that the laws of chemistry represent an independent, new context, one that *cannot* be explained in reference to the mechanical motions of the smallest particles. That is because chemical and mechanical conditions of atoms are mutually exclusive, in the sense that the state of an atom in which we know its chemical behavior cannot be described in terms of the mechanical motions of that atom's building blocks. Conversely, more accurate knowledge of the electrons' mechanical behavior renders knowledge of chemical properties impossible. The earlier notion that Bohr's theory attributes chemical behavior or, more generally, the sensory properties of matter, to motions of the smallest particles, has to be understood in this sense: Modern atomic theory has clarified in every detail how chemical conditions fit in with the mechanical or, more generally, with the previously known physical laws. At the same time, atomic theory has given the laws of chemistry their most likely final mathematical form.

The laws of chemistry must mesh so flawlessly with physical (i.e. mechanical, electrical, optical) natural laws that no contradictions can arise even in the most remote conclusions we draw. Formulated mathematically this would be expressed as follows: The matters addressed by classical physics, that is, those related *directly* to objective processes in space and time, are contained as restricted cases in the more general cases of quantum theory. Seen from this formal viewpoint, the domain of classical physics appears as a special case, as a part of the general domain of the laws established in quantum theory. Yet, conversely, it is classical physics that creates the prerequisites for the coherent formulation of quantum theory in the first place. For its laws also must apply indirectly to objective processes in space and time, since in the area of material events one can only speak meaningfully about objective processes in that sense. In a certain way, classical physics provides an integrating component of language with the help of which the contexts of quantum theory can be put into words at all. That is why one may perhaps compare the scientifically ordered understanding of the world that has developed over centuries to language-learning in children. At first, concepts are formed from simple experiences; they do not actually represent reality but idealizations of it. Those simple concepts are the prerequisites for understanding more complicated things and for formulating more complicated concepts or idealizations. These latter concepts provide the possibility to speak at all about more intricate matters and, hence, to determine the limits to the applicability of the simple concepts.

Thus, the reciprocal relation between chemical matters and the more specifically physical ones manifests itself to us as follows: So that laws of chemistry can be understood in their relation to those of physics, a broadening of the framework of classical physics into that of quantum theory is required. The laws of quantum theory are superordinate to those of classical physics, incorporating them as a restricted case. In addition, the laws of quantum theory contain the laws of chemistry and, more general, the whole body of laws related to the sensory properties of matter. The range of phenomena of quantum theory allows us subsequently to determine the boundaries of the two domains: physics and chemistry.

(c) The Boundaries of the Domains

It was long held that Newtonian physics describes precisely the domain of reality about which we can speak with the help of the concepts of "mass," "location," "velocity." This raises the question how such an assertion can still be defended *in light* of the laws of quantum theory. Clearly, one could draw conclusions from the validity of classical mechanics, regarding, say an atom's electrons, that do not prove correct in experience.

Quantum theory solves this seeming contradiction as follows: The laws of mechanics can generally be unambiguously applied to a system only when its characteristic determinants (such as the location and velocity of each point of its mass) are known exactly. According to quantum theory, these determinants exist partly in "complementary" relations, which means that exact knowledge of one of them excludes exact knowledge of another. Thus, complete knowledge of a system cannot be achieved in the manner proposed by Newtonian physics. What can be achieved is knowledge of a "state" in the quantum theoretical sense and that means that, as long as we are talking only about mechanical concepts, we may have knowledge of the determinants that is *imprecise* in some measure. Wherever a practically *unambiguous* conclusion is drawn from that imprecise knowledge about certain properties of the system according to the rules of classical physics, the conclusion is legitimate also within quantum theory and in experience. This is the precise meaning of the statement that Newtonian mechanics exactly represents the domain of reality which can be described in terms like mass, location, velocity. In cases where an unambiguous conclusion is not possible, however, classical physics, too, is left with statistical statements as its sole recourse. Such statistical statements of classical mechanics prove to be incorrect both in experience and in the quantum theory that represents that experience. Classical mechanics cannot provide information about the *frequency* of mechanical formations. In the small-scale atomic domain, that frequency is determined by conditions of a different kind that are alien to mechanics, and that can to a certain extent fill the gaps created by the imprecise knowledge of the classical determinants.

To cite an example from atomic physics: If it is known that an electron moves around a proton (a hydrogen nucleus) at a distance in the order of one ten millionth of a millimeter and at a speed in the proximity of one hundredth of the speed of light, classical mechanics would conclude that the electron possesses some sort of energy that is on a comparable scale to the electron's binding energy in the basic state of the hydrogen atom. It is also given that in a large range of energy all values of energy would be about equally probable. In reality, one may conclude with a high degree of probability from this presupposed knowledge that we are dealing with an electron in the ground state of the hydrogen atom, that is, an electron of very *specific* energy. Hence, the *statistical* conclusion drawn by classical mechanics fails in light of experience because it ignores the existence of non-mechanical conditions. In this case, the conclusion forgets about the chemical forces that give rise to a very specific binding-energy of the electron.

This complete failure of classical mechanics in determining the frequency of atomic systems may be seen as the very first proof that there are non-mechanical structures of law at work in nature. The existence of stable atoms that have very specific, constantly unchanging properties (binding-energies, chemical forces, colors, etc.) and are comparable in number to the overall number of atomic systems is something beyond comprehension for classical mechanics, proving that there are contexts of a different kind.

On the other hand, re-examining the laws of classical physics in studies of the individual atom, wherever an unambiguous examination is possible, will accurately confirm the validity of those laws.

When judged as to how directly they relate to objective processes in space and time, the structures of law that govern physics and chemistry show themselves to be of *equal* order and subordinated to the totality of quantum-theoretical contexts. Nonetheless, a certain ranking of these two groups of laws becomes apparent when one studies the degree of generality which their activity in nature manifests. For while the action of the laws of classical physics is always clearly discernible wherever material processes are going on, the properly "chemical" processes are tied to more special external conditions. Given our present knowledge, no chemical processes in their actual sense could occur, for example, in a star's inner matter, glowing red-hot in enormous heat and under huge pressure. Those total and often sudden alterations in the structure of matter that may arise when different substances are mixed together–such as in combustion, crystallization, solution–and which we call "chemical" processes, cannot possibly occur in the presence of extreme temperatures and pressures. That is because there, atoms that are no longer stable at all, and thus no longer fuse here into molecules or into solid or fluid matter. At most one may call them "chemical" processes at most in relation to atomic nuclei, but these are not processes one would call "chemical" in the usual sense of the term. Some experiences in astronomy suggest that the state of the world some five billion years ago–something already nearly beyond our conceptual abilities–was such that the universe's entire matter was concentrated in a relatively small space and glowing in extreme temperatures. We must assume as a result that the chemical course of events began to develop only gradually when matter, conglomerated now into special stars, would have cooled off at their surfaces. Similarly, life must have begun only at a much later stage and under much more specialized conditions on one or perhaps many stars. We are forced to conclude: While the laws of chemistry–or the laws of quantum theory superordinate to those of physics and chemistry–are perpetually" valid" (otherwise one could not speak at all of a "law"), the special processes that we call chemical emerged only gradually as the universe developed.

Viewed from a different perspective, the structure of laws governing chemistry may be regarded as a "higher" organizational form of matter that is tied to more special conditions: In order to occur at all, chemical processes require pieces of matter that exceed a specific minimal size. In any case, an individual molecule represents the bottom limit for a chemical process, for an individual molecule, the actual elementary particle, can no longer undergo any chemical process as it has neither valence forces nor binding energies or any other chemical properties. But one cannot establish *exactly* the bottom limit of the size of the pieces of matter that are used in the observation of chemical processes. As soon as one approaches the bottom limit, the concept of "chemical process" becomes questionable. Only when arbitrary and artificial definitions are provided to give this concept a meaning more precise than what is warranted that limit can be set more accurately. What applies to the boundaries of life applies similarly to those of chemistry: In the area of smallest organisms the question of whether a certain object is a "living being" or a piece of "dead matter" can be answered also only in terms of arbitrary definitions.

The difficulty of separating chemical processes in the field of minute units of matter from purely mechanical processes implies also that the domain of chemical phenomena cannot be taken to be closed in the same sense as the domain of classical physics. Rather, chemistry continuously crosses over into mechanics and electricity with which, however, it can be re-united only when we let go of the demand that we only {ever} think in terms of processes in space and time and, instead, acknowledge the connections of quantum theory. It seems, therefore, correct to regard quantum theory as the next higher domain, superordinate to classical physics, while regarding chemistry as a special projection only of that domain onto the level of objective space-time processes.

Classical physics represents that idealization of reality in which only objective material processes in space and time are spoken of, independently of the question how those processes can be determined for example. Quantum theory encompasses a broader domain of reality; it can be viewed as the idealization wherein a state is described by indicating the degree of probability that certain material, space-time processes will occur when made accessible to observation (by means of external intervention.) Hence, this is the sort of idealization in which reality appears at every moment as a certain multitude of possibilities for objective realization.

(d) Chance

Just as all recognition and representation, and hence all language rests on repetition, that is, on the possibility of discovering something "like" among diverse circumstances, so too scientific ordering of the world begins with repetition, in the orderliness of law. Generally speaking, even the attempt to represent something "objective" in language is based on the presupposition, itself having been justified by success, that a solid chain of cause and effect leads from "the object" to ourselves and, as we act, from ourselves to the object. For without that solid chain of cause and effect, we could not draw inferences from a "perception" to a specific "process" and every agreed upon explanation of what is taking place would be without foundation.

Classical physics does justice to this situation to the extent that from the outset it combines the representation of objective space-time processes with the presupposition that those processes are completely determined. Its model pictures spatial-material systems that are sealed off from the external world and whose temporal course is determined by their present state for all time.

Contrary to this idealization, quantum physics' concept of state creates an entirely new situation in relation to the question of determinacy of natural processes. In place of the *closed* system as something that happens in space and time, there is the totality of possible space-time processes that take place when a system is *under observation*, that is to say, when it is in *connection* with the external world. Here one could expect complete determinacy at best only when, in addition to the state of the system, the details of the intervention necessary for the observation could also be taken as given. But then accurate knowledge of those details could itself be attained only by precisely observing the means of observation that cause the intervention if the procedure of observing were itself not again dependent on an intervention that cannot be monitored. In other words, one is drawn into an unending regression which precludes meaningfully raising the issue of the determinacy of the processes of nature.

Thus, in the domain of reality whose structural connections quantum theory depicts the laws of nature do not lead to a complete determination of what happens in space and time. Rather, what takes place is left to the play of chance (within the range of frequency determined by those connections.). In this domain, chance may initially be understood as "meaningless"; that is how Goethe too perceived the word chance in the section of his Theory of Color referred to. After all, the word "meaning" envisages a direct relation to us as thinking and suffering beings about whom we cannot yet speak here, where we are occupied with laws of nature.

Nor can we assume that that the events seemingly left here to the play of chance were themselves fixed by natural laws of a different kind or of a higher order. For that would mean that the frequency of space-time processes would, under given conditions of quantum theory, be different in certain circumstances from what one would expect according to the rules of quantum theory and that would suggest that these rules do not yet represent the correct laws of nature. But this is improbable given the many accurate confirmations of those rules. And yet, even this question also is seen in a different light when one considers that perhaps there are systems or more accurately: that there is knowledge of systems to which quantum theory's concept of state can no longer be applied. Obviously, such systems would no longer be bound by what quantum theory declares regarding probability and could, therefore, be subsumed under quite different sets of connections. In this sense, and in this sense only, one may say that today physics leaves open the possibility that certain processes that appear to follow the play of chance in light of nature's laws known to us, are perhaps determined by connections of a higher order.

The answer to that question must in any given instance be decided according to one's experience. Let us look at an example that helps to judge well the scope of this issue and the possibility of its being settled: The formation of crystals, on the one hand, and of living organisms, on the other. The laws of nature established by atomic physics help us understand that crystals are formed when liquid matter congeals in the course of cooling off. From atomic laws one can deduce not only the fact that atoms arrange themselves in rank and file into solid material, but also the particular way the crystal is organized, its symmetries and structure. But according to the laws known to us, the unique external form of the individual crystal is left to the play of chance. Even if we could replicate exactly the same external conditions for a crystal's formation, the form of the emerged crystal would still not always be the same: The drop of water, cooled off in the cold air, congeals into the snow crystal. Unless there is external interference, the crystal's symmetry will always be that of a hexagon. But the unique form of the small crystal-star is not predetermined by any law of nature. Within the limits imposed its hexagonal symmetry, the size of the droplet, the way it cools off, etc., chance designs the infinitely varied patterns of tiny stars and plates that seem to us just as artistic as a kaleidoscope's multitude of images.

In this example experience, too, yields no grounds for believing that, let us say, the formation of snow crystals is tied by higher webs of connections to very specific forms. We may therefore believe that it is the play of chance even though we are not fundamentally *forced* to draw such a conclusion, for we cannot claim that we know the quantum-theoretical state of the droplet of water before and during the formation of crystals. Assuming that the laws of quantum theory are correct, one is constrained to acknowledge that chance is at work but only in such examples where the quantum-theoretical state is known with certainty. Take, for example, some radioactive material of which one definitely knows that practically all atomic nuclei are in their normal state. Here the emission of radioactive particles *must* be left (within the quantum-theoretical range of frequency) to the play of chance if the laws of quantum theory are truly legitimate.

Even if we believe that the growth of an individual crystal was not predetermined, so that another, somewhat different one could have developed just as easily, nothing has been decided about whether the coincidence to which the crystal owes its form was "meaningless." For a crystal's formation is an irreversible historical act which as such can also play an important role in the context of our lives or that of the world, even if it was not predetermined. Contexts of a kind that justify us to use the word "meaning" can also be connected to events which could just as well have transpired differently without any reason whatsoever.

When comparing the formation of crystals to the origin of living organisms, one encounters, upon closer examination, a completely different situation (superficial analogies notwithstanding.) Even though the laws of atomic physics can probably give us a complete understanding, right down to the last detail, of the complicated chemical compounds that make up the structure of organisms, from the perspective of atomic physics a living being still seems an enormously improbable arrangement of atoms. And even if such an arrangement—one we don't know at all precisely—were at some point assumed to be given and the question arose as to how such a system evolves over time in its interaction with the environment, atomic physics would likely predict a sequence of changes the course of which ordinary language would eventually come to speak of as death and decomposition. In any case, there is no aspect known to us of the laws of quantum theory that could offer any explanation at all for the formation of organisms. The borderline territory between biology, chemistry and atomic physics has not yet been sufficiently researched to exclude with certainty the assumption that the laws of organic life derive from those of atomic physics. But many researchers tend to believe that the laws of organic life are of "contexts of a different kind" that are not already contained within those of atomic physics. The "forces" of nature that are capable of forming crystals would thus indeed be able to form the complicated chemical compounds from which the organism originates, but the pure play of those forces and of chance do not suffice to render the organism as a whole comprehensible. It would mean—just as Goethe assumed in his Order of Reality—that the laws of organic life actually belong to the next higher domain of reality, whose relation to quantum theory is similar to that of quantum theory to classical physics. This view, first stated in precise terms by Bohr, is to be the starting point for the following observations concerning organic life.

4. Organic Life

In order to understand chemical phenomena, science followed two paths. On the one hand, the chemical transformations of matter were depicted as objective processes whose regularities were studied, yielding a description meticulous to the last detail of those phenomena and their connections. But here they encountered the dilemma that in the domain of the smallest particles of matter one cannot draw a clear boundary between chemical and electro-mechanical processes. This implies that there can be no closed formulation of the chemical phenomena. On the other hand, it is precisely the connection between chemical processes and motion, that is, the play of forces between those smallest particles, that researchers chose as their starting point for understanding. Thus, the exact formulation of chemical laws was achieved in some instances at the expense of the ideal of the "objectification" of events.

Similar aims are being pursued on two paths to understand organic life. On the one hand, there is the meticulous description of living processes and, on the other, the analysis of those processes in reference to the interconnections of life with physical-chemical processes. One may perhaps speak of yet a third way, unlike any in lower domains because we are living beings ourselves and because consciousness seems intimately related to the forces that hold the organism together as a single entity. The question must remain open for now to what extent one can speak about organic life at all without any direct reference to consciousness. Perhaps the domain of reality that encompasses organic life cannot be delimited from that other broader domain that includes, to the extent that it is accessible to ordinary language, the knowledge of the human soul and the space of which is filled by the flood of unconscious processes below the ebb and flow of consciousness. But for the time being we shall not speak of this side of the problem of life since what we are to address now is the delimitation of organic life from the next lower domain, that of physics and chemistry.

Only in the last decades has science accessed the border zone between biology, physics and chemistry. For now, we cannot claim to have an understanding of life's processes comparable, for example, to what quantum theory has achieved for chemical processes. That is why as yet no solution can be offered to the basic problems related to the delimitation of life-processes from physics and chemistry and that are raised from perspectives such as vitalism, materialism, etc. Attempts—no matter how frequent—to solve these problems through reflections of a general nature, such as those of epistemology, or through generalizations of specific experiences, can at best throw light on the questions raised from a certain position and bring to light a small part of the truth. But, on the whole, the toilsome path of research cannot be skipped over; perhaps we must await the results of individual experiments of many future decades before we have a relatively clear picture of the relation of life's processes to those of physics and chemistry.

If one intends to develop an overall map of the scientific order of reality, one must be clear from the outset that the material presuppositions are only sufficient for a precise implementation of this plan in the lower domains up to that of chemistry. Similarly, our image of the higher domains, beginning with biology of our time, cannot be drawn much more precisely than, for example, the way the Greeks mapped the lands [C37] beyond the Euphrates at the time of Alexander.

What this reflection attempts is to sketch out at least the contours of that overall map. But as far as biology is concerned, it too must rest content with raising the known basic problems and discussing the possibilities for their solution. Bohr's thesis may be [C38] of help here. According to him, the delimitation of biological matters from those of physics and chemistry may be conceived of similarly as the delimitation of chemistry from physics. For it may signal a way to think about understanding that delimitation without major contradictions being raised by experience and scientific conscience. In any case, Bohr's supposition allows us to offer clear answers to the basic questions related to delimitation which may prove useful as working hypotheses to the further advance of research.

(a) The Relation Between Biological and Physical-Chemical Laws of Nature

When studying organic life, our attention automatically focuses on two characteristic features observable in a similar way in both the most highly developed and the most primitive living beings. On the one hand, a living organism behaves completely differently from an entity we refer to as "dead matter." A living organism performs various functions such as metabolism, reproduction, etc.; its development manifests a peculiar stability vis-à-vis all outside disturbances and its behavior supports in numerous details the impression that it is oriented toward specific goals, that it serves an end planned by the organism. With every living being we feel a certain relationship with ourselves. But, on the other hand, that living being can obviously also be looked upon as a physical-chemical system; that is, it may be compared to a complicated machine. For wherever their course may be observed in detail—one can always trace the various ways an organism behaves back to physical-chemical processes. But we know of no process yet which shows that physical-chemical processes follow different laws in living beings than in dead matter.

This curious situation is presented in two natural, but opposing theses concerning the relation of biological laws to those of chemistry and physics:

Vitalism proposes that in the living organism there are still "forces" or holistic structures at work that are not present in dead matter and that cannot be regarded as a result of physical-chemical dimensions. This life-force, said to be the only thing that really distinguishes the organism from dead matter, was in earlier times thought of as most likely a force in the ordinary sense, present in the living being in addition to the forces of physics and chemistry. But, given the experiences of modern biology, we are more inclined to replace the concept of life-force, signaling as it does an all-too-close analogy to the concept of force used in physics, with one that emphasizes the characteristic behavior of the living being as a unity over against the physical-chemical behavior of its parts. This is how Driesch, for example, uses [C39] the concept of entelechy or unity, maintaining that the physical-chemical process is not to be drawn in its details into this superordinate dimension but that, nonetheless, "the whole is more than the sum of its parts."

This vitalistic thesis is opposed by another that was upheld primarily in the age of materialism. Perhaps it can now be stated as follows: the laws of physics and chemistry have unlimited validity in relation to the organism and, for that reason, also exclusively determine its behavior. This second thesis can claim in its support that thus far no deviations from the known physical-chemical laws have ever been observed in the organism and that in most cases where an organism's form of behavior was at first taken to be typically non-physical, it could later be construed as a physical-chemical model of the process in question. Familiar manifestations from the world of living organisms come to mind:

Flowers turn their leaves and blossoms toward the light. This clearly is a case of purpose-related behavior in line with the living being as a whole. We come to understand it much more through analogies to our own wishes than through causal chains of physics and chemistry. Nonetheless, it can be shown that this movement toward the light is caused by certain photochemical reactions in the cellular web that lead in normal chemical ways to an expansion of the cells and to movement.

Driesch referred particularly to the following process: The eggs of intricately formed living beings, such as the sea urchin, can still be divided during their earliest phases of development without losing their developmental abilities. The two halves both develop into complete living beings. Such behavior, too, allows the mistaken impression that the game of cause and effect, played by physical and chemical forces, could hardly explain this phenomenon and that it would rather indicate that the organism's entire map is drawn, as it were, in the cell and that it achieves its ends despite even the crudest external interventions. Yet here too one may point out that similar processes repeat themselves externally in the formation of crystals: a drop of water, falling to earth through the cold atmosphere, solidifies into a snow crystal. But even if the drop were divided into two parts beforehand by external intervention, each part would once again have to form a complete snow crystal. The laws of physics and chemistry completely explain this process.

Now, the first thing an assessment of the pros and cons of vitalism must decide is whether there are internal contradictions in the assumption that the "higher" aspects of organic life are qualitatively different from the physical-chemical ones while not at all affecting the "lower" aspects. In other words, it must be determined whether we are dealing simply with laws of nature "above and beyond" those of physics and chemistry. This understanding may have been guided by the idea that it is the entelechy, the image of the whole, imprinted in the organism, which determines its behavior, guides its physical-chemical processes, something like an engineer controls and directs an engine that in itself functions according to the simple laws of cause and effect. But such a comparison necessarily leads to the assumption that as currently formulated the physical-chemical laws do *not* apply everywhere to the organism. For the engineer can direct the engine only because he is himself body and matter and intervenes in the course of events in a purely physical manner. For example, he must move a lever and in so doing work physically on the machine. And this means, for example, that the total energy is not conserved for the machine alone but only for the system of machine and engineer together. The comparison of entelechy to the engineer in control necessarily leads to entelechy being seen as a physical force. Thus, the principle of the conservation of energy is applicable only to the system of body and entelechy together, while the total energy stored in the body alone would not necessarily be conserved. The laws of physics can be kept in their present form only if they are expanded into laws that somehow also comprise the force-fields of entelechy [C40]. Perhaps one cannot yet refute such a conception conclusively because the balance of energy in the body, for example, cannot be determined with very great precision. But it is only a small step from the conception just described to the assumption, surely long since abandoned, for example that the weight of a living being changes at the moment when the soul leaves the body at death. Such views have been left behind, not because they could be conclusively refuted but because we have come to see that the problem of the relation of biological to physical-chemical dimensions would not be solved at all, but only postponed by the introduction of an additional force-field, irrespective of whether it is called life-force or entelechy. For one, it would require a study of how such a force-field assists in determining the processes related to the organism as a *unity*. For another, entelechy itself would have to be accessible to the methods of examination used by physics. But, thus far, those methods have failed to find a trace of such force-fields anywhere. As it stands in any case, all experiences to date give every reason to believe that there is no such life-force in that sense.

The above comparison might change, say, if it were assumed that the entelechy directs physical-chemical processes just as the engineer's *mind* directs the engine. This image may well very accurately represent the real situations but the relation of the mind to the body is just as problematic as that of the truly organic aspects to the physical-chemical processes. Thus, the comparison provides no answer to the question whether one may assume without internal contradictions that the biological laws of nature do not really follow from the physical-chemical ones but that, in spite of that, they leave their functioning entirely untouched.

With respect to the numerous attempts to subordinate the physical-chemical laws under the laws of biology as a system of connections without violating the laws of physics and chemistry, it must at least be emphasized that the physical and chemical laws *wholly* determine the behavior of a system the material properties of which are known. If the organism is a purely physical object in the sense that we can believe that the quantum-theoretical "condition" of this system, composed of many atoms, is fully known, then its further behavior is fixed by the laws of quantum-theory; there is no room for superordinate biological laws. This state of affairs is not changed even by the fact that generally the quantum theoretical condition determines further behavior only statistically, that is to say, it indicates only in how many similar cases a certain result will occur. For a superordinate biological dimension would change precisely the frequency of an event—for example—in the sense that events which protect the organism against external disturbances would take place in preference over others. Therefore, quantum theory, like classical physics, leaves no room for such superordinate dimensions.

If one wants to make room for properly biological phenomena that are not simply the result of physical-chemical phenomena—many experiences support the creation of such room—one may with Bohr call to mind the relation between quantum theory, chemistry and classical physics. The following comparison suggests itself: "Entelechy" or "the holistic structure" "guides" the physical-chemical process in the organism similarly to how the field of matter-waves "guides" the movement of the electrical elementary particles in the atom.

This comparison is protected first of all against the objections that were raised against the earlier comparisons: The field of matter waves is not a force-field that "acts upon" matter but is in a certain sense another aspect of matter itself. The principle of conservation of energy applies to electrons and their electric interactions as precisely as can be verified. For its formulation it is neither necessary nor possible to introduce, in addition to the electrons, the field of the matter-waves. After all, the guidance of electrons by the field of matter occurs in a different mode: the building block of reality that we call electron is not only, or not always, a minute elementary particle that moves in space and time according to the laws of classical physics. Rather, it has that property only in experiments where we examine its spatial location. This building block "electron" may in different situations also be a wave-process and, as such, is subject to the laws of wave propagation. As quantum mechanics demonstrates in detail, it is only here that stable atoms can emerge to interact with chemical forces. If we apply the substance of these remarks *mutatis mutandis* to living organisms, we will come to the following conclusion:

Living substance is not only, or not always, a material structure constructed of atoms that changes according to the laws of physics and chemistry (or, more generally, according to those of quantum-theory.) It has that property only (and also always) in experiments where we examine its physical-chemical behavior. But living substance can be something else in other cases, for example, an organic entity; as such, it follows biological laws. That alone allows for the emergence of stable organisms that can also enter among themselves into the relations characteristic of organisms.

Just as the laws of chemistry can be conceived of as a certain projection of the general laws of quantum-theory into the area of objective processes in space and time, so the laws of biology can also be regarded as the projection of the nexus of the next higher domain of reality into that area.

The meaning of Bohr's hypothesis becomes apparent only when one examines in detail his conclusions as they relate to biological processes.

What follows from this thesis, in the first place, is that physical-chemical laws are said to apply absolutely in the organism: Wherever mechanical or chemical changes are observed in an organism, it must be possible to explain these changes in terms of physical or chemical effects—unless say, the examination of the forces brought in to bring about the process necessarily introduces significant changes to the process itself. Therefore—using an example from genetics—when in the fertilized ovum chromosomes line up in pairs so that a paternal and a maternal chromosome each unites with the other to become a pair, the uniting process must be capable of physical explanation either in terms of, for example, electrical forces, chemical affinities or similar effects. Alternatively, it must turn out that a physical examination of the forces bringing about the uniting process cannot be carried out because such an examination would decisively disrupt the process itself. We cannot determine as yet with complete certainty which alternative applies in this specific case but the first is more likely since, compared to the sensitivity of biological processes, the uniting of chromosomes is for all intents and purposes still a relatively crude process. Therefore, in this case it should be possible to follow the physical-chemical causal chain several more links until one perhaps finally arrives at the point where the experimental verification disrupts and changes the process itself. It is not improbable that the examination will finally end at such a point; otherwise it might perhaps not be possible to understand the overwhelming frequency of the processes that manifest purposefulness of the organism as a whole.

This already signals the other side of Bohr's thesis: He claims that the laws of biology are not merely a consequence of those of physics and chemistry just as chemical laws are not to be understood as, say, a consequence of the mechanical and electrical dimensions in the atom (as classical physics sees them). But the fact that classical physics falls short in explaining the stability of atoms could only be proven once we had very precise knowledge about the details of the structure of the atom. Similarly, we will only be able to provide a compelling proof that physics and chemistry fall short in explaining the structuring of organisms when we have much more precise knowledge of the structure of the most minute organisms than we do now, perhaps only when the study of biology is "complete." For the time being, we must content ourselves with this observation: From the point of view of present atomic theory, the organism appears to be an utterly improbable formation, somewhat like the improbability, according to classical mechanics, that countless atoms of the same kind would exist in a single substance.

So that the laws of biology do not contradict those of physics and chemistry they are, according to Bohr's thesis, to be understood as being in a similar relation of complementarity to the latter as the laws of chemistry are to those of mechanics. This means that the observation: that an object is a living cell relates *exclusively* to the exact knowledge of its quantum theoretical condition (and a fortiori to the knowledge of the location and velocity of its elementary particles.) When we know that what is before us is a living cell, that knowledge provides information about a series of properties of that formation that we could probably *not* deduce from its physical behavior alone. But, on the other hand, in this situation perhaps one will have to do without knowing its quantum theoretical condition. Nevertheless, there is always the possibility of examining its physical condition experimentally. However, such an examination would—presumably—require such a strong intervention that the life of the cell would be destroyed in the process if it is to produce a truly accurate determination. Of course, it is possible to vary the level of intervention from very weak to very strong. Many biological experiments are designed to provide information about relatively coarse features of the physical or chemical behavior of the cell and thus, may be satisfied with only slight intervention. Perhaps such an insignificantly disrupted organism will allow the typically "biological" features to recede a bit in its behavior and, in so doing, follow in every case the laws of physics and chemistry insofar as can be determined. The traces of the intervention may perhaps have already faded after a short time. But in other cases—the x-ray photography of large protein molecules [C41] comes to mind—the intervention required for the examination will result in decisive changes in the object under study.

Thus what Bohr's thesis presents is this: The knowledge that we are dealing with a living organism creates a situation that cannot be expressed solely in terms of the concepts of prevailing physics and chemistry as we have known them. The laws of biology represent a separate dimension that cannot be directly combined at all on the level of space-time processes with the laws of physics and chemistry. The task of biology: to clarify the integration of biological laws of nature and the physical behavior of matter (in particular of atomic matter) has therefore to lead—similarly as in the case of quantum theory—away from the level of objective processes in space and time. Only then can one achieve an overview of the next higher domain of reality that also embraces life. That domain may be called the biological domain even though biology in fact represents only a projection of that domain onto the level of objective processes.

(b) The Structure of the Biological Domain

Chemical laws could not be formulated exactly, nor could the question of the nature of chemical forces be answered as long as the focus was restricted to actual chemistry, that is to say, to the qualitative transformation of measurable quantities of substances. Only with the advance into the chemistry of smallest quantities of matter (atoms and molecules), into the boundary area where chemical and mechanical processes can no longer be sharply distinguished, was it possible to discover and exactly formulate the natural laws that encompass chemistry and mechanics simultaneously.

Similarly, as long as we restrict ourselves to the study of life forms that are visible to us, we are not likely to succeed in formulating biological laws exactly or answering the question of the nature of the forces that decisively shape life. Only as we move on to the area of the minutest of organisms, the boundary area where living begins [C42] cannot be sharply distinguished from large molecules, may it become possible to track down the natural laws that encompass biology, physics and chemistry at one and the same time. This boundary area has become open to research only in the last decades; the road to the discovery of natural laws is likely to be yet a long one. The question must for the time being remain open of how the logical freedom created by Bohr's thesis may be exploited by a positive assertion about the biological dimension. Nonetheless, we shall address at least some possibilities concerning the substance of this prospect.

In order to make some judgment about the biological dimension it is important from the start to state that the origin of life or, more precisely, of living organisms on the slowly cooling surface of the earth was a unique process, one that we cannot reproduce experimentally. That is why the application of the concept of "law" to life-processes is problematic, since a natural law is by its very nature a statement about processes that can be repeated as often as one wishes. To be sure, it is possible to think about the earth's cooling as something that happened any number of times, but if we wanted to say what happens in such a cooling process we would leave behind the area of experimentally ascertainable statements. At best a comparison, made at some future time, of the formation of organic entities in different stellar systems could replace the repetition of that process. However, for the time being we do not know whether such organic formations exist on other bodies in the universe. In any case, biology for now refers to organic life on our earth. Laws can be therefore be spoken about in as many places as there are repeatable biological experiments. But these experiments always presuppose the existence of life as it has developed. The *origin* of living organisms from inorganic matter remains a unique historical process.

One could now, for example, take Darwin's principle of selection as the key to understand the processes of life, as many researches have done. This view assumes that as temperature sank on the surface of the earth and it became physically possible for larger organic molecules to form, more complicated formations developed of their own accord. Some of them— or perhaps at first only one of them—had the characteristic that they were capable of producing others of like structure from chemically related matter. According to Darwin, the formations most suitable to such processes then spread out, reduplicating themselves at the expense of less suitable ones. The process of reproduction also occasionally led by coincidence to somewhat changed formations (mutation). The most suitable among these changed forms prevailed so that finally ever more complicated beings came into existence ("practiced", as it were, by nature). The more developed among these manifested ever more clearly many characteristics that may be considered as particularly "useful" in the struggle for the preservation of the species. If one adopts this Darwinian position, the defensibility of which can be determined only on the basis of experience and which likely has not been determined in every detail, then saying that a certain cell "lives" would be equivalent to saying that here is a formation nature "practiced" and shaped over the course of a billion year-long development. Knowing that one has to do with such a formation would, in the sense of Bohr's thesis, be complementary to knowing its exact physical condition. Thus, this situation is similar to the theory of heat, where knowing the temperature also is information of a different kind from knowing the aspects of mechanical definition. However, it would in principle always be possible to ascertain the physical condition at the same time as one ascertains "life."

The position just described, which is most closely related to mechanics, may be summed up like this: In order to understand the processes of life it is basically only necessary to add the concept of historical development to the concepts of physics and chemistry. Above, and contrasting physics and chemistry, biology would then be characterized as the science which not only tries to establish conceptually what is basically repeatable in nature but which also holds among its basic presuppositions the unique development of our earth.

But surely it is debatable whether this view really depicts in an unconstrained manner our knowledge of the processes of life. For even if one could explicate in terms of the principle of selection all the manifestations of organisms usually cited as arguments in favor of vitalism, the very existence of consciousness surely shows that there are features of life-processes [C43] that certainly cannot be explained by the principle of selection. For while it may perhaps be conceded that living beings, whose reactions are centrally coordinated, fare especially well in the struggles of life, that signals nothing about whether this reaction-center is "given" to the individual living being in the form of consciousness. So, if we already know that the complete description of life-processes includes concepts, such as consciousness, that can in no way be traced back to physical-chemical concepts, it seems most natural to assume that the simpler biological and physiological concepts, especially that of "life" itself, are strange, new in comparison to the physical-chemical ones. For, certainly, the concepts of biology are given to and understood by us quite separately from the physical-chemical nexus. Of course, nature's relationships as described by biological concepts must align themselves without contradiction to the lower orders of phenomena, and the notion of selection will play a significant role in the analysis of that alignment. But the biological concepts are completely independent and are formulated by us because we ourselves are alive. And when it comes to differentiating the boundaries between the two domain-areas more concepts will be needed than only that of selection.

If one gives Bohr's thesis this broader interpretation, one must assume that it will at some later time be possible to formulate the phenomena of biology as expressions of natural laws, despite the uniqueness of the earth's history. The idea just won't go away that on some other celestial bodies, to the extent that their physical conditions are similar to those of our earth, beings must have developed that would have to be described as living beings, and that therefore the biological nexus of such bodies is also one subject to basically repeatable processes. Of course, we cannot for the time being decide to what extent one can justifiably imagine a later phase of biology, say, in terms of the model of quantum theory. But for now one will be guided by such analogies and, therefore, must ask what general statements can be derived from them about the interconnections of that third domain.

The first thing to be observed here is that the laws we are looking for and their action cannot be limited to the living substance but that they must be about quite general connections that touch all that happens and—as follows from the concept of law—are generally binding. The origin of living beings is perceived then as merely a special effect of those laws; this is comparable to the existence of stable atoms and molecules as a special effect of quantum-laws.

A particular conclusion to be drawn from this is that in the field of most minute organisms no sharp distinction can be made between living and dead matter. One can decide with complete certainty in relation to a large formed body whether it lives or not. But in relation to the smallest organic formations the concepts that usually help us make that decision fail so that artificial definitions are needed to maintain that distinction.

When seen from this perspective, the well-known thesis that "life would be able to originate only from life" appears in a different light. We do know for certain that the more highly developed organisms emerged in earlier times from more primitive organisms; hence, the thesis naturally leads back to the most primitive, single-cell living beings that must have been present at the beginning of this development. But the question whether those most minute living beings derive from living or dead matter cannot be settled. One may express this by saying that living matter is all there is at all; or one may also use the term primordial conception (*Urzeugung*) for the gradual development of larger and more complicated organisms. But this should not be associated with the probably erroneous idea that the behavior of the matter used in the formation of microorganisms can be completely described in terms of the concepts of physics and chemistry available to us to date.

This next level of the order of science may perhaps distinguish itself again from quantum theory by a further expansion of the concept of state. The "state" we may be able to come to know on that level will perhaps be to an even greater extent a sum of possibilities and to an even lesser extent an objectifiable reality in space and time. It is highly likely that an even more important role will be played in particular by that peculiar "reaching-over" in space and time that manifests itself, for example, in quantum theory namely that an electron can be a small particle and a wave. This apparent "reaching-over" of a body into distant space-time areas comes about, as we know, in that the body manifests itself, so to speak, only as specialization of a more general connection. In quantum theory specialization results from the intervention needed for the observation that informs us about the location of the body. No one can as yet predict what conditions will look like in detail in relation to the range of living phenomena. Much of what we know, for example about biological instinct-actions, (such as the migratory patterns of birds) renders plausible the idea that biological phenomena, which are not marked off in elementary ways by the spatial-temporal expansion of bodily organisms and their forces, can be operative in large space-time areas. The very concept of the biological "function" of an organ is, after all, not directly tied to a specific material condition in space and time. In accordance with the analogy of quantum theory, it would once again be more accurate not to speak of effects upon large space-time distances, but of phenomena that cannot be described simply as "effects" which produce the image of such seeming effects when projected into space and time.

Examples of the "reaching-over" in space and time have often been cited. Larvae of many species of insects engage in extensive preparation before their pupation; they construct so-called cocoons whose intended purpose is realized only *after* pupation in the living being's changed condition. Here we cannot speak of experience, learning or reflection. The event being described can be depicted without constraint only if one considers the entire development of the form from the egg to the death of the fully formed insect as a holistic event including all "instinct-actions" and independent of the particularities of the interaction with an environment often marked by chance. In any case, what such examples show is that the material process, which often varies much in its details, does not play the decisive role for the biological event—just as for example, the mechanical movement of the individual electrons does not determine the formation of the crystal in the process of crystallization. What is much more determinative in that formation is the existence of a situation which evades mechanical description, namely Bohr's "stationary condition" where one cannot speak of the electrons' movement. Perhaps the existence of certain biological functions (metabolism, reproduction, etc.) may have to provide in a similar way the actual basis for an understanding of life's processes. The study of the physical-chemical properties of microorganisms yields only knowledge of the processes nature plays with in order to materialize those basic biological forms. On the other hand, the biological nexus does have to "line up" with the physical-chemical laws so that no contradictions arise anywhere. This alone already shows us that an infinite quantity of material leading to an understanding of the biological dimension can be gained from the study of the physical-chemical properties of the smallest living beings.

As has been alluded to several times, another characteristic feature of the level of the order of reality following that of quantum theory consists perhaps in that the formulation of the appropriate laws is impacted by the fact that we ourselves are living beings. A number of diverse reasons attest to the likelihood of this:

For one, the laws determining the genesis of organisms as well as every interaction between an organism and the "environment" and, hence, every observation of an organism, must be operative. And so, surely, the fact that these laws apply just as much to what does the observing as to what is being observed will enter the conclusive formulation of those laws. Similarly, it is essential in the formulation of quantum theory that its laws apply not only to the atomic system to be observed but also within the observing apparatus.

But, secondly, there may well also be relations above and beyond this between living organisms that cannot be traced back at all to the "objective" concepts of classical physics.

The novelty in quantum-theory's epistemological situation consisted in the understanding that we can observe only what cannot really be separated from us so that the concept of "objective observation" becomes in a sense contradictory.

Life places us once again before a new epistemological moment; the analogy of this problem to the problem of quantum theory discussed thus far was not intended to be interpreted as suggesting that the solution of the epistemological difficulties of the problem of life must lie in the discussion of the intervention associated with every observation. What needs to be stressed is, in Bohr's words, that "life is not an experiment of physics" [C44]. What may well be the most essential aspect of this new epistemological situation is that the very concept of "observation" contains for a living being features that cannot be defined in physical–objective terms. We can enter into a direct relation to a living being, a relation that is neither capable of nor in need of analysis in physical concepts. Our bond with the beings that are most closely related to us, namely other human beings, becomes most clearly apparent to us; a child comprehends a mother's demeanor, for example, long before language enters in. But such bonds also connect us—consciously or unconsciously—with the more highly developed animals. For example, the inner uncertainty that overtakes us in face of another living being's death-struggle is a clear sign of the close bond among all that lives. We may leave open the question how much of such connectedness is still felt in our relation to the lowest organisms, such as the single-cell beings. But it is only because we ourselves are alive that we can even determine with any significance that we are dealing with a living being. In like manner, the word "love" can also be understood only by someone who has already been met with love. Only a misunderstanding of the epistemological situation discussed here could lead to the misleading idea that much could be gained by defining, that is, going back to other concepts. We know without explanation what these words mean. To be sure, in biological science even the word "life" too must undergo the process of "defining concepts ever more precisely" that we described earlier. But the fact that we know what life is because we ourselves are alive is part of the epistemological situation of the domain of reality that encompasses life.

This direct relation to all that lives that we have been talking about appears, moreover, to be only a special case of the more general interconnections characteristic of the phenomenon of life as a whole. We sought to express this by noting that the biological function is primary vis-à-vis the objective material process. For example, the following biological processes are obviously closely related internally and differ in some ways only by degree: the construction of an organ from many cells of one kind; the formation of an ant colony that lives according to specific laws and reacts holistically almost like an organ; the coming together of human beings in a community. In each case, the same image of the whole hovers here, as it were, over the whole into which the individual members unite, over the development of the emerging organic unity.

We must also consider in this context the close connection of a living being with its material environment. A merely superficial observation can distinguish between the living being's body and the dead material environment. On closer inspection these boundaries turn out to be imprecise. Does the nourishment just taken, the air breathed, belong to the body or not? From what moment in time are they to be regarded as part of the body? Does the snail's shell belong to the organism of the snail or the web to that of the spider? Would not also the earth in which the plant puts down roots have to be seen to a certain extent as part of its living nexus? And, moving directly to one of the last links of this chain:

We human beings, too, encounter a landscape not only as spectators. On the contrary, every act of coming to know something is connected with an unconscious probing of possible, living relationships that can grow between that landscape and us. These relationships can become so close that their violent rupture can lead to severe disruption of an entire course of life. Touching a living being is something different from touching a dead object. Even upon entering into a landscape we can sometimes clearly feel the "forces" that flow into us from there.

The nexus of this third domain of reality that embraces life is often designated as a "higher" form of the organization of matter, higher than the nexus of physics and chemistry.

The fact that life can generally develop only under very special external conditions, specific temperatures, degrees of humidity, etc., may serve as an indicator for that designation, as does the acknowledgement that organic formations which manifest the typical marks of life appear only above a certain minimal size. At least several million atoms seem to be needed to form a "living being." Finally, the existence of death is also characteristic for that higher organization, that is, the almost fitful and never reversible dissolution of the being's inner organic unity. This dissolution is caused and accompanied by certain demonstrable material changes in the body of the living being. But death's onset also has a side that cannot be described simply in terms of material structure; it may be made clearer by the notion that the image of the whole, which in life relates the parts of the organism one to the other, fades at the moment of death and loses its bonding power.

Such a cessation of biological interconnections may also be observed where it is not a matter of an individual's life coming to a close. Thus, too, for example, the relationship of two human beings that enters our consciousness as love, can come to an end and be subject to the same fate as the organic unity of a living being is subject to at death.

To be sure, in all the cases in which until now, when looking at life we have talked of *human* life, we will have correctly described only part of the nexus. For in the case of the human being, the existence of consciousness and the participation in a higher mental order of things fundamentally changes every individual process.

(c) The Unique Position of the Human Being

What would need saying here is that we human beings are not only bodies and living beings, but are partakers in a higher order, that in a certain sense we designate that place in nature where the highest orders that give shape to matter become manifest.

Here we seek to emphasize, on the one hand, the special place of the human race in the history of the earth's development and, on the other, its ability to understand mental structures.

There can be no doubt that in the course of the earth's history the human being developed from other organisms which were more primitive in their external construction. It may also be taken for granted that humankind's early ancestors had great similarity with some lower species of animals living today. Nonetheless, it does not follow that those species would also be capable in principle of giving rise once again to a comparable series of developments. Rather, it seems that the development of species frequently went the way of "specialization." According to most biologists today that development goes forward in such a way that mutations simply follow the rule of chance and that then selection from among the given conditions determines the further course of development of those most fit for life. That selection obviously results in special capabilities being enhanced to high perfection but, because of this, no further development is possible since the organism has by then been encapsulated too much in the specific nature of its life struggle. Specialization prevents the actual development into a higher form. However the process of the organisms' formation played itself out in detail, it is clear that humankind is today's living representative of the central line of development. If one does not want to assume that development has just now come to its completion, one must concede that today's human beings too are in the act of changing very slowly into more differentiated, still more highly organized beings. Perhaps one may even compare past and future as closely as to say that present humankind is the origin of diverse lines of development that lead partly to other specialized groups of organisms and partly to more highly organized living beings capable of further formation. It is true, such development takes place in time-spans of extraordinary lengths in comparison to the short span of the life of an individual, a people or a cultural community. One may therefore rightly ask what sense there is in considering such a distant future at all. But our age today is inclined not to think of a relatively imminent and anticipated end of the world; instead, it is more apt to compare future possibilities with what happened in the most remote, yet still perceivable past. Development and its course over billions of years before us therefore forces us to take up the idea of its future continuation. This ever so remote past and future–as little as we have to do with them now-in a certain way provides the frame within which we see the picture of our contemporary world history.

The idea of further development taking place in time-spans beyond our imagination gives rise to another conclusion as well. In its first weeks of existence, a human embryo, while externally bearing the features of earlier stages of development, participates unconsciously in that higher nexus which embodies itself in the human being alone. One has to assume, for that reason, that the reality of today's human being does not exhaust itself in the configuration we have come to call the biological, psychological or spiritual order and that, instead, the human being fundamentally participates also in every higher configuration, the activities of which become manifest only in much later epochs. This "participation" can refer initially only to the basic possibilities of development. But it may also refer beyond that, to the fact that in the lives of today's human beings future formations already announce themselves. But we cannot speak about that development, given the nature of the matter. After all, a language can only be formed for that part of reality where our life takes place. What is at issue here may be summed up in the following somewhat unclear statement: In contrast to all other living beings on our earth, the human being is the only one possessing access to the creative forces.

5. Consciousness

Presumably, no one today dares venture as yet into delineating the higher domains of reality or determining how they are mutually related. However much has been said and thought about this part of the world, studies of it have almost always had to restrict themselves to describing and ordering that which can be experienced. Only a few attempts have been made to enter an almost impenetrable darkness and access the area where these domains of reality connect one with another and with those of lower order.

(a) Consciousness and Biology

People at all times seem to have agreed that the existence of consciousness, of conscious life, may be regarded as the next higher stage of reality, the one superimposed as an umbrella, so to speak, above that of organic life. Given that the connections between an individual's life and her or his consciousness are so close, one has to ask whether one may meaningfully separate consciousness and life into two domains at all. On the contrary, there are many indications that these are effects of the same kind which relate the parts of an organism to a common, integrated whole and which may manifest themselves in consciousness as wishes or feelings, impressions or acts of the will.

In response to this question people have often noted—as in the studies of Carus [C45] or of modern psychiatry—that the soul's conscious life draws consistently on a much more extensive unconscious life. One may compare this to the play of waves on the ocean's surface in relation to the movement of the sea below them. Every observation of the processes within our own consciousness informs us that only a small part of our thoughts enter into the bright light of our consciousness. Another, larger part hurriedly transverses a kind of semi-dark space while the largest part of the processes manifest themselves, when one attempts after the fact to identify them, as an indeterminate movement on the shadow side of consciousness. The idea that there is a continuous flow of conscious processes into an utterly "unconscious" area imposes itself irrefutably on us.

But if this is true, another question arises right away: Are the unconscious processes not directly identical with life as such, that is to say, with the processes that may be regarded simply as the manifestation of organic unity, the way a living being sets itself apart from the environment? Carus asked this already. His response distinguished between a "general" and a "partial" unconscious. For him, the former is identical with the action of the formative forces that give shape to the life in question. The latter already belongs in a certain way to that very soul which also reveals itself in consciousness. It is, so to speak, the darkness into which the beam of consciousness can shine. Carus described in detail the relation of these phases of the unconscious and conscious life that constantly permeate one another.

Now, even though it appears certain that we are able to describe a constant flow from the fully conscious processes of the mind to the wholly unconscious action of the organic creative forces, it leaves wide open the question as to whether the actual *cognitive situation* of observing consciousness is not fundamentally different from that of observing life. We have come across something like that several times before: It seems that even though in nature only fluid processes take place, the conceptualizations we approach nature with, the language that we use, necessarily cause sharp boundaries to appear between the diverse domains of reality. Chemical processes in most minute spatial domains, for example, do indeed continuously merge into the movements of the elementary particles (atoms and their electrons.) Nonetheless, chemical changes are conceptually distinguished so precisely from the processes of the elementary particles' motion that, as quantum theory teaches, the ways of conceptualizing these two phenomena are in an exclusive relation of complementarity.

Similarly, one may safely assume that a new cognitive situation prevails when we are looking at the soul as a unity that receives and feels impressions, that entertains desires and makes decisions. A sharp delineation appears to be clearly discernible here between two domains of reality. But this delineation does not really relate at foremost to two discrete domains of reality and conceptually ordered. Rather, it is once again related only to their projection onto the level of consistently objective events though not always on to that of time and space. Only insofar as the soul (our own or that of other living beings) is the *object* of observation must it be distinguished fundamentally from the totality of the biological range of connections that comprise the living being as a whole. In the observation of processes relating to the soul, there probably is an increased level of difficulties similar to those already noted in relation to the examination of atomic processes: a substantial portion of events connected with the soul will elude objective determination to a certain extent because the act of determination itself decisively affects those events. However justified it is to regard events related to the soul as objective, and however nonsensical it would be to subordinate soul-reality as derivative or secondary to material reality, it is equally necessary to emphasize that the objectification of mental processes signifies a particularly extensive idealization of actual reality. Memory capable, for example, which lets the same thoughts pass through the consciousness again in the course of firming a soul-process, can make sure that the part of the process which took place in the bright light of consciousness is repeated with reasonable accuracy. What memory is certainly not able to achieve is to assure that the greater portion of the unconscious processes associated with the firming of a mental process will take the same course in the new situation of attempted fixation as it had before. What we are dealing with here is a general characteristic of scientific method: In the fluid nexus of nature, it is a fundamental, given that processes flow somewhat differently than where we isolate and examine them up close by means of experimentation or mental analysis. To be sure, every verbalization of a process is already a form of isolating and scrutinizing it!

Thus, insofar as they are the fixed object of our reflection, processes of the psyche are part of a specific domain of reality and can probably not be related in unambiguously determinable ways to processes considered to be the direct expression of biological functions. But this does not rule out that the connection between the purely biological processes and those of the soul is nonetheless so close that the latter may be taken simply as the form that the biological course takes in consciousness.

Thus here, among other issues, we also raise the problem of the so-called psycho-physical parallelism. At first glance, it seems reasonable to believe that, parallel to every sequence of thoughts, there is a certain electro-chemical process in the brain where thoughts are logically connected and by association, but that the electro-chemical processes are determined by physical-chemical laws according to cause and effect. But for a variety of reasons, such a statement is superficial. The examination of the differentials of electric potential in the brain—insofar as it can be undertaken without significantly disturbing the train of thought—may well lead to the conclusion that a certain chain of ideas will, under certain conditions and in the same individual, always also result in the potential following the same temporal process. This is the truth in the statement just mentioned. However, what must be noted here is, first, that this process of potential can in no way be the same for different individuals, different ages, etc. For example, if significant changes have taken place in the brain as a result of injury, the process of potential associated with the same idea would look completely different from what it did before. Secondly, that formulation obviously overlooks the fact that ideas are not given to us with our birth like our organs and their assigned functions, but that we come to them only in togetherness with other human beings. It is not at all wrong to state that the performances of the same organ in two different living beings are essentially "the same." But one can talk about the sameness of two living beings' *ideas* only when an understanding about the idea's content can be reached through language or gesture. Just as a certain biological function is often realized materially in quite different ways, the same idea can also be subject to different material and biological processes. For that reason it is no longer surprising that a logical chain of ideas "runs parallel" to the electro-chemical processes in the brain for which concepts such as "logical sequence," etc. do not exist. After all, drawing logical conclusions is something learned, something implanted on the brain only through communication with other human beings and through experience.

As has already been indicated, the question of the strict correlation between ideas and the brain's electro-chemical process must in the end seriously consider the fact that every examination of ideas or what goes on in the brain interferes more or less significantly with the process to be examined. Every electrical intervention on the brain will give rise to ideas, acts of the will, body movements just as every search for ideas must cause electrical and chemical processes in the brain. But it is not the task of these lines to pursue such complex connections; we need only to remind ourselves of the fundamental cognitive difficulties we face in asking about the relations between physical and biological events.

(b) Consciousness and Reality

However much the attempt to objectify the processes of the soul makes them appear as something fundamentally different from biological processes, they would seem to be correctly described if we look at the whole spectrum of the soul's processes as the form of the connections existing in our consciousness that are indicated by the concept "biological functions."

Thus the manner in which our environment is mirrored in our consciousness is also a direct expression of the biological relations connecting us with that environment. For this reason our study naturally turns back to the peculiar changes in reality we spoke of at the beginning of this essay. Our ability to adapt is greatest in childhood, the world we are born into has the greatest influence upon us and changes us the most and our relation to the world is not yet fixed by the formation of special abilities. At that stage, our reality is borne also by the relations that are in the process of formation between ourselves and the world immediately around us. Just as every relation has to be mutual, the power to change and to give form to reality is greatest at this stage as well. Only as this growth process approaches its conclusion, does the ability for specific, intentional living connections mature. Only then can we grow into a landscape or enter into relationship with another human being. Moreover, the emergence of such a living connection is so sudden, so much something that happens to us without notice and as if by a higher power, that it can fill us with something like a deep and holy terror. It is as if the divine itself had come down to earth and were speaking to us through just that human being or landscape. What takes place here in detail is something that only the poet can describe in parables, for no one who has encountered God at this place would dare talk in ordinary language about that event. But what is certain is that behind the reality of such a living relationship the entire rest of the world taken in by the senses loses its force for a long time, either by receding into the shadows or by being enfolded into and taking part in the radiance that now floods all of consciousness.

Looked at from this perspective, one may understand the other, opposite event described at the beginning, where the relation of a human being with the surrounding world seems to change from a living encounter into a rigid, mechanical connection. What apparently occurs here is a real destruction of living connections, a process akin to death that can come about through a catastrophe in the life of that human being's soul or, perhaps, also through the gradual weakening of the organism.

We have to emphasize another, particular consequence of becoming aware of the living connections, namely, that, contrary to all lower connections, consciousness leads to a decisive separation of the individual from her or his environment. A crystal can be broken down into separate parts or fused with other crystals into a larger one; in every case, whatever emerges from these actions, each of these formations always retains the properties of crystals. Similarly, in the process of fertilization two cells may become one and subsequently divide again into two identical cells; the division can be continued almost at will. Thus, the cell too undergoes changes in the course of time which can prevent it from always being recognized as the same individual cell. Consciousness alone radically sets apart a specific unit; it is simply not imaginable that one individual's consciousness could fuse with that of another or that a division of consciousness would occur. It is obvious that such setting apart of an individual "I" from the rest of the world is possible externally only because already in the organic world, the individual living being represents a separate unit, to a certain extent set apart from the environment. That is why one cannot speak of consciousness in relation to lower forms of life where the division of one organism can produce several more. (One thinks of the case of plants where a cutting from a shoot, planted in the ground, can often become a new plant.) But it would not be correct to say that this setting apart of the individual "I" happens only *to the degree* that the biological placement of the living being allows. Rather, consciousness always makes that separation {always} with utter precision; the situation through which we come to understand this allows for no gradual transitions. By its nature, the "I" is an indissoluble unity; it may come into or go out of existence but cannot be subject to the process of division or fusion. Thus, whereas in the lower domains of reality processes follow one upon the other in colorful, yet regular succession, so that either by cause and effect or by coincidence one situation leads to another, the higher domains are marked by the existence of ultimately unchangeable entities that, just because they cannot become something "other," can only come into and then go out of existence.

One may raise the fundamental objection here that the entity of which we are speaking as strictly separate of its environment exists only conceptually, for this total separation is forced upon us only when we attempt to objectify and describe the structure we call consciousness. But we must keep in mind that scientific language after all must consistently objectify and describe when it seeks to represent a domain of reality.

What we said earlier about how the diverse domains of reality must necessarily fit together most likely also compels us to conclude that the unity represented by a particular human beings' consciousness must have ceased to exist before the body itself moves toward its dissolution in death. If nonetheless we assert that the human soul can continue to live after death, what we most likely refer to is above all the experience that the structure a person has imposed on her or his environment is able to go on working and can be directly felt in that person's environment. When, let us say, we enter the rooms where someone close to us has lived and that bear everywhere the marks of her or his being, we may have the impression that the person's spirit is still alive in those rooms; and this spirit can have an extremely powerful influence on our action and life. Of course, it can be said that this is in fact only the effect of a reconstruction from material traces, which becomes especially apparent here and guides our action. But who knows whether the "explanation" of the spirit's ongoing activity in terms of the concept of reconstruction is actually better than the explanation of biological processes in terms of the physical-chemical processes in the organism. Perhaps the force active here is once again something holistic that cannot be reduced without coercion into a sum of elements that can be reconstructed and is therefore something cognitively other. But whoever would attempt to depict the connections that impinge on our consciousness through that force would again enter into the field that can be spoken of only in parables.

6. Symbol and Gestalt

At the gate that leads from the domain of simple consciousness to the field of the mind and everything connected to it stands the "symbol." However little we know for now about this array of connections, it may even be justified to gather up in this word symbol the entire domain of reality which is to be depicted as a cohesive configuration above and beyond simple consciousness. For everything of the mind, whether in language, science or art, rests on the use of symbols and their power. Matters of the mind are not tied to bodies but are transmitted through symbols. At the same time, the symbolic power of an object or a process is something entirely objective, comparable to consciousness or life; more correctly perhaps, symbolic power is something objectifiable. It is a reality no weaker than, let us say, the reality of consciousness or that of a biological configuration. Were one unwilling to look upon the power of symbols as having any more than secondary status, one would distort the picture of this aspect of reality. Just as there is not even a hint of the existence of consciousness in any of reality's lower domains, so the fact that things, movements or sounds can "mean" something, cannot be anticipated from phenomena of a more elementary kind.

What comes to mind here is the flower that to us is the image of life and of youth, the embodiment of perfect, self-contained beauty: the rose. The relation of the human being to this flower is not based in any of the levels of reality discussed hitherto. The strictly biological relation is not particularly close, to say the least: as a plant, the rose is neither related nor of any use to the human being. As an organism, it is as perfect in itself, no more and no less, than the thistle or the wood louse. It has no substances precious to us. But for us human beings, it is more than all that can be captured in biological or physical-chemical concepts. The shimmer of its hues that radiates unclouded from its blossoms, the breath of wind that carries its scent to us, touch the innermost part of our soul. That is presumably as much an objective state of affairs as any of natural science. But, of course, its true substance can be expressed only in parables. For in its original form, the symbol stands close to the central domain of the creative forces: symbol "signifies" nothing definite, nothing that can be articulated; it does not mean to direct our thought in a particular direction. Rather, it transports us into a particular state, makes us receptive and opens doors to domains of reality that are difficult to access. It is the task of poets to speak about this. What has to be said here in relation to the example of the rose can perhaps be expressed in the words of Gottfried Keller's Ode: Seeing the rose can make the silver string break forth with song; for it is, after all, its profound sound that makes life important to us.

We can even list reasons for the existence of symbols and for the special meaning the rose has acquired for us. But this does not imply that this level of reality could be explained in terms of other levels or be traced back to them. There are reasons as well for the existence of consciousness as a purposeful summary of the organism's response to the external world. Nonetheless, consciousness is something different at the level of cognition than the biological functions. The reasons, which up to a certain extent can explain the higher level of reality on the basis of the lower, merely prove that the different domains of reality "match one another"—as can be studied clearly in the relation between chemistry and the mechanics of the movements of electrons.

When we try to name and to deal cognitively with the particular effects the rose has upon the human soul, these effects may take on a specific form. As it opens up, the rose bud becomes for us emblematic for youth, the white rose [C46] for purity. But with this specialization the symbol already begins to move away gradually from the domain of the creative forces. It steers our thoughts in a specific direction. When this process of specification and narrowing is pushed further, one eventually arrives at the large group of specialized symbols that are the precondition for all understanding and, concomitantly, all thinking: language and writing.

(a) The Means of Communication

The road from the pure, living symbol to the firmly fixed term of recognition is marked by infinitely diverse intermediate forms that permit an initial level of communication among human beings. This includes, above all, gestures; more original than language and writing, they are able to communicate ideas and emotional content from one human being to another.

Whereas we have to learn language and writing, the medium of gestures is ours from onward birth: Amazement, joy, rejection communicate themselves much more directly in the child's face than in the adult's. We see already in the animal crying out in terror the warning sign and the call for help.

But gestures alone only allow for communication of a very general kind. Only a continuously progressing specialization of such utterances, eventually leading to utterances with increasingly precise "meanings", brings about the kind of communication in every detail that constitutes the precondition for all life of the mind: Language and writing.

From the perspective of the history of evolution, the forebears of humankind will be defined as human beings in the real sense of that word only from the time when they had language at their disposal. Here, not unlike the development of an organ, the development of language and, concomitantly, of rational thinking, is the exploitation and fixation of a possibility that was present in earlier stages of development. It is conceivable that the development of this organ called "thinking" is now complete and that, therefore, future, more highly differentiated beings will not differ fundamentally in their ability to think rationally from today's human beings. Based on this assumption the continuation of development would have to proceed, as is always the case in organic life, from the central domain. One could conclude that the places where we are dealing with living, not yet finalized symbols, namely poetry and art, already suggest the direction of further development. But such vague conjectures should better be left unspoken.

The bond established between human beings by language and writing is of a whole different kind from those of the biological area. While we can quite suddenly be overtaken by that special communion with other people, a communion that is rooted in life itself, the bond created through language is something we can deliberately establish and dissolve. Thus, this bond is subject to our consciousness and our will.

The access that language and rational thinking create to the rest of the world obviously has, in turn, an exclusive, complementary relation to the relation that the biological area has to this environment. Migratory birds find their route South precisely because they cann*ot* think or talk about how they get there. We human beings are not born with the capability to find such routes as a matter of course; instead, we have the possibility to find the way by means of maps once we have learned how from other people. And it would likely not only be due to the economy of nature that it did not endow *one* living being with *both* capabilities. Instead, the relationship to the environment that emerges from the step by step advancement of rational thinking does not allow that this environment is, as it were, automatically given to us in its entirety.

That is why, for example, where life itself brings us into relationships with other human beings there arise the tensions that we speak of as a struggle between thinking and passion.

Here, too, lie the roots of the difficulties much talked about these days that unleash a veritable war of opinion with the phrase "the mind's[1] hostility toward life. [C47]" In face of this feud, one should always keep in mind that the possibilities offered by the increasingly refined use of specialized symbols may perhaps indeed become exhausted, but that further development can only originate from the level of reality where there are in fact symbols. For life in itself is dim and only the power to create symbols and to understand them turns us from living beings into human beings.

The complementary relation between the biological interconnectedness and rational communication is also clearly apparent in the diverse ways by which we reach the biological or the mind level of reality at the beginning of life. Drives are innate; we would have them even if in the course of our development we never came in contact with other human beings. The mind level, on the contrary, opens itself to us only through interaction with other human beings who already possess this reality. The development over a span of a billion years from the single-cell being to the human being is replicated, in a certain sense, in the development of every single individual in the period between the fertilization of the cell and birth. Humankind's spiritual development, by contrast, which has evolved from the first attempts at communication until today, is repeated by each individual through her or his interaction with other human beings during the first year of life in what we call "learning." What may be ours by birth perhaps is the "substance" that can be set alight by the flame of the spirit and by certain basic patterns of thinking, but not the spirit itself. In saying this, there can hardly be any doubt—if we use the word substance in this figurative manner—that we may have been endowed with a more or less suitable substance. For there are indeed human beings who are quite primitive or who have been ravaged by hereditary illness whose lives are barely able to be brightened by matters of the mind and spirit at all. What seems highly problematic, however, is the question whether or in what manner *specific* ability of the mind is innate.

---

[1] The German word *Geist* challenges the translator in that it cannot consistently be translated as *mind*. There are times when the more accurate rendition has to be *spirit* (the "s" in lower case) or *Spirit* (upper case "S"). Another translation is *intellect*. The context determines which term is used in each case. (Trans. note.)

The fact that the life of the mind is handed on from generation to generation by language, that it is to a certain extent distributed among human beings and not made to reside in the individual person, this is quite likely what makes Bohr question whether specific spiritual abilities, such as musical talent, can be biologically inherited [C48] at all. Bohr makes the following comparison: While it is possible to construct better or worse aircraft that can fly short or long distances, it is not possible to construct aircraft that can fly better from Berlin to Stuttgart than to Munich. But it seems to me, staying with this example, one may object that very different types of aircraft can be constructed: those that fly long distances or climb to great heights, cargo planes that carry huge loads, or those designed to reach the highest possible speeds in races. Depending on the terrain and the weather, one aircraft will be more suitable for one route than for another. Or, to steer the discussion back to our subject at hand, the ability to take in what is spiritual can hardly be measured on *one* scale only. That ability is more likely comprised of many diverse components that, while we may not yet recognize them at all by their various kinds, they nevertheless, whatever their kind, enter into every special domain of the spiritual in different degrees. That is why I think it probable that, independent of experience, talent, even specific talent, can be inherited. On the other hand, the life of the mind is transmitted, after all, only through human interaction and language, for which reason human beings' environment has to be of decisive import for their spiritual development.

In assessing to what extent a human being's particular kind of spirit and character may be conditioned by biological heredity, we also need to keep in mind that what we are used to calling someone's "kind" is comprised of many components. The features that are purely drive-related, that have nothing at all to do with the spiritual dimension, likely make up a considerable portion of these components. In addition, where we ordinarily speak of "heredity," we do not primarily think of mere bodily heredity as depicted in Mendelian laws; rather, in light of the fact that almost always children grow up with their parents in their early years, the "heredity" we refer to here is from the outset one of blood *and* interaction. (After all, interaction too is part of the biological process.) Perhaps for most practical purposes it will not be all important to distinguish two components which almost always impact on life in tandem. For example, when people talk of how musical talent may be inherited, surely what they generally have in mind is this gift's being handed on to the next generation through the organism *as well as* through the interaction that enriches the spirit. But presumably the question of how the talents of the spirit are passed on (that is, of the *precondition* for the spirit's enlightenment) through biological inheritance can in principle be raised and answered.

However that question is answered, the layer of reality borne by symbols is certainly separated cognitively from all other lower layers, especially the biological ones, by an abyss. References to an "effect" of biological phenomena on the spiritual area are to be understood only in the sense of what we called the necessary "matching" of the different domains of reality. This is similar to how we may speak about mechanical effects in the reciprocal interaction of atoms or of chemical effects in the organism.

In the domain of the mind it is also obvious, even more so than in, let us say, that of biology, we are dealing with configurations that cannot simply be related to space and time or ordered in space and time. Thoughts can leap over space and time and we can transport ourselves in our imagination into times past or yet to come and into distant places. An ordering in terms of space and time is possible only in relation to the carrier of the thought, for example, the person who had the thought or the book it is written in. But such an order does not contain what is essential. Ordering thoughts according to their substance or their logical connections is incomparably more important. In seeking to establish a spatial-temporal order of thoughts according to who or what carried them, one comes once again to the old questions philosophers have pondered ever so often: What is the location of the soul in the human body? Is it not present everywhere in the body, seeing that we are able to sense and act with all of our body's parts, etc.? Questions like that help us understand that in order to fit the processes of the mind in very specific ways into space and time, we must do violence to them.

It becomes apparent at this point that even the most consistent attempt to objectify a layer of reality does not always result in a conceptual system that could be attached without difficulty to that of classical physics. The qualities of matter that chemistry deals with scientifically do indeed still fit without constraint into the classical view of the world ordered according to space and time. It is questions such as: What color or temperature does an electron have? that finally remind us that there are limits to this conceptual system of qualities. But the distance between diverse conceptual systems becomes more apparent already when we look at biology. For example, the familiar questions concerning the spatial boundaries of an organism or the "forces" that hold together a colony of ants point to the difficulties that arise when one tries to establish a connection here to a simple space-time order—something not at all hard to understand in light of the epistemological situation of quantum theory.

In this context, let us touch briefly on the old question of whether our "knowledge" of the inner world of the mind is more certain than our knowledge of the "external" world that is occasionally falsified as a result of sensory deceptions. Today's prevailing opinion is that we can no longer assign statements of priority according to their degree of certainty to this or that layer of reality. A statement such as "I think" is initially no more certain than "I write," for a deception of the imagination could have been at play, for example, in both cases. But a general statement that is not based on a specific experience, such as the assertion: "I exist," may well be as certain as any correct mathematical statement. But it may also not be more certain, that is to say, the words "I" and "exist" can be so defined, so axiomatically fixed, that the statement "I exist" actually becomes correct. Similarly, the axioms of arithmetic also are fixed in such a way that, let us say, the statement $2 \times 2 = 4$ turns out to be correct. However correct the statement "I exist" is in itself, it says nothing as yet about the extent to which we can make use of the concepts "I" and "exist" so that we may find our way in reality. We know equally little to what extent we can make use of the concept of numbers in ordering our experience. The basic law of Newtonian mechanics: "force = mass x acceleration," too, applies with the same degree of certainty as the statement "I exist." For in both cases certainty is based first in the conviction gained through experience that we have captured a trait of reality in such a statement and, second, in the knowledge that this statement derives from a system of axioms which, in itself free of contradictions, as a whole depicts particular traits of reality. And more than that can most likely not be achieved at all in the attempt to speak about reality.

Thus even though it can be readily understood that, in terms of the degree of certainty, there is no difference between the two statements: "cogito, ergo sum" and "energy = mass x acceleration," it must be emphasized that they occupy completely different epistemological positions. "Cogito, ergo sum" is the foundation of reflective thinking; it is the a priori precondition for reflection being possible in the first place. "Energy = mass x acceleration," on the other hand, is only one of classical physics basic axioms.

The question of the a priori validity of the forms of apperception: space and time, discussed again and again since Kant, is also appropriate to raise at this point. Together with the validity of Euclidian geometry, the independence of space and time, these forms of apperception, as they have been handed down to us, have proven themselves in the interaction of human beings with the world and owe their validity precisely to that fact of having proven true over time. Granted, those forms have now become *more* than empirical givens because, as Kant rightly emphasizes, they in turn create the very precondition that allows us to have experiences at all. But the fact that we cannot have experiences in other than in these forms of apperception does not support the assumption that those forms remain unchanging in perpetuity. Rather, the very existence of the theory of relativity signals that here, too, we can and must relearn; it is conceivable that generations to come will gather their experiences differently from the outset and order them other than we do. Biologist Konrad Lorenz maintained that the forms of apperception had to be understood as "innate patterns" comparable to the instinctive actions of animals. On the one hand, that view explains why those forms create for us the necessary precondition for all experience; we can imagine, of course, that there are no things in space but not that space and time do not exist or are different. On the other hand, "having been proven useful" in reality establishes those forms as conditions for our existence; only beings whose forms of apperception conform well to their environment can prevail in the struggle for existence. That is why over the course of millennia that struggle can bring about changes of those forms of apperception in human beings along with changes in the environment. We may be able to reflect on such changes rationally but cannot implement them directly. Thus Lorenz' assumption that the a priori forms of apperception and categories are innate patterns in the biological sense, steers a middle course that is just right between the extreme view that the forms and categories are unconditionally valid independent of experience and its opposite, namely that they are strictly a matter of our experience. Accordingly, the process whereby the forms of apperception conform to the environment is a learning process not of an individual human being but of humankind as a whole. It is not a process of, let us say, a large number of human beings at a given time coming to new insight into a new context. Rather, it is one in which those who do not gain such insight die out and the others who have a "talent" for this new understanding or to whom it has been given from the start make better progress.

Looking at this point again at the question of how the layer of reality pertaining to symbols is related to the layer of organic life, it becomes apparent that the delimitation cannot be drawn quite as simply as Bohr tried to. The fact that the life of the mind is transmitted from one generation to another by language, that, in other words, it is in a way distributed among human beings rather than located in the individual human being, does not warrant the assertion that what is of the spirit cannot be passed on through heredity. Rather, the question here has to be put the other way around, as follows: "What are the basic elements or preconditions of the mind that can be passed on genetically?" In view of the manifold and complicated "innate patterns" in the animal world, the existence of such hereditary basic elements becomes probable from the outset. Surely it is a first step in the right direction to count the forms of apperception and categories that constitute the very precondition for all experience among those basic inheritable elements. One may surely assume, for example, that human beings growing up from birth without language and without all the specialized symbols nonetheless experience the world of space and time and that these are independent patterns of order for all external experiences for them as well. But this by no means answers Bohr's question of how the capability for the gifts of the mind are inherited, except that now it is no longer a question of principle but becomes one of detail. In answering it, the issue is to determine how the biological domain and that of the spirit's processes are interconnected in such a way so that no contradictions arise. Only in the lowermost domains, those of physics and chemistry, has the matching of the diverse layers of reality been fully recognized; but there is no reason we should not assume that we will be able to have a clear view of this matching someday in relation to the higher domains as well.

If the a priori concepts are said to be dependent on the biological history of development as we have just described it, a misunderstanding may result: The objective course of events in the world and, in this context, the course of evolution in particular, is taken to be the primary process; and the way the world is reflected in the mind only follows from that process. In other words, it would appear as if the "existence of an objective, real external world" had to be deduced from the above described dependency on the biological domain. But that is certainly not what is meant here. After all, today's natural science has been forced by its epistemological experiences straightaway to address the question what meaning the assertion of "the existence of the objective, real external world" may have. It can probably be no more than a more cautious statement that a large segment of the world of our experience can be successfully objectified. This more cautious formulation no longer leads one to an over-emphasis on the "objective" world. The dependency of a priori concepts on evolution discussed here is meant to stress that it must also be possible to discuss a priori concepts within the context of the biological sphere and that one must be able to assign those concepts a place within that domain as well. How that order matches with the orders of the other layers of reality is another question which we will not pursue here any further.

(b) Art

How can what is of the mind in symbols be handed on other than by specializing their "meaning?" This question poses itself when attention turns from the specialized symbols back to their original, living changeable form. One possibility for grasping their substance is through the *ordering* of symbols. This is the basic process on which all art is based and which elicited much admiration as early as the Pythagoreans: If it is at all suited to this process, almost any material sensory input can trigger processes of the soul, can become the bearer of spiritual or, more accurately, artistic substance when it impacts us in an ordered form, for example, according to established, mathematically comprehensible structures.

This basic process is exploited in the most primitive way in the kaleidoscope in order to design a pretty pattern from an accidental arrangement of colored glass. Through multiple reflection, two mirrors positioned at an oblique angle, manage to create from such an accidental arrangement a regular hexagon filling a circle for example. In most cases, this simplest hexagonal symmetry is enough to produce a pattern that can create the impression of a small work of art. Now if a single mathematical symmetry on such a simple scale suffices to give meaning to a multicolored medley, how much wider the range of possibility available to artists to express the mind's substance in the manifold structures and symmetries that they are able to impose on their materials. But only in the rarest cases do beholders become aware of those structures. It is likely that in many cases becoming aware of the symmetries hinders rather than furthers the perception of the spiritual substance that they preserve.

Ever since the Pythagoreans, the harmonies in music have been the most noted example of this. Strings of equal tension, whose lengths are kept in simple rational ratios, produce a harmonious sound as they vibrate together. Thus we obviously seem unconsciously able to perceive measure and number in those sounds, and the rational ratios of the number of vibrations may indicate a basic structure which, together with other structures, allows sounds to become music. Generally speaking, music is the most striking example of the phenomenon that symbols which mean nothing individually may become carriers of spiritual substance by being ordered. In all other forms of art, a special meaning of the individual symbol almost always resonates along with the individual meaning: in poetry the ordinary meaning of the words, in painting the object envisaged. But in music it is almost senseless to speak of a special meaning of individual sounds; apart from the actual structures, at best general qualities of perception, such as loud-soft, calm-agitated, etc., come into play.

Music is produced by very different kinds of structures. In addition to the rational ratios of the number of vibrations, we could list tempo in the sequence of sounds (beat, rhythm, phrasing, that is, the rational ratios of how long a sound or group of sounds are held), symmetries in the presentation of the melody or theme (repetition, mirroring, abbreviation, doubling, etc.), symmetries in the sequence of harmonies ("cadence"), repetition, mirroring, etc., in joining different voices ("polyphony"), movement structure (repetition and grouping of themes in the individual "movement", reciprocal references among different movements), etc. Only this fullness of diverse structures creates the richness of music that lifts it above the play of patterns that we see, for example, in the kaleidoscope. It is this richness that helps us understand that a musical score can convey the bliss that a particular time experienced to later generations for centuries to come.

One of the aspects of the layer of reality whose boundaries we can indicate with the concepts of "mind" or "symbol" is that we can pass on the things of the spirit from one human being to another in the way symbols are ordered. What we cannot do is to "explain" this transmission in terms of conceptions that constitute the lower layers of reality. Nonetheless, we cannot consider this fact utterly independently of such connections. It already follows from what we said earlier about the different orders of reality "matching" that the order of symbols can hand down spiritual substance only when it is accessible directly to our senses, that is, without the detour through rational thinking. The rational ratio of the number of vibrations two sounds produce may be experienced as harmonious because there is in our auditory organs a special and characteristic condition that corresponds to that ratio that differs significantly from the condition of vibrations in somewhat divergent, non-rational ratios. But, when it comes, for example, to the interaction of two colors whose optical frequencies are in a rational ratio to each other, they will *not* create the impression of any kind of harmony. That is because the condition of our eyes and our optic nerves is not essentially different as we simultaneously behold colors of rational or neighboring non-rational vibrations.

Thus, if spiritual content is to be passed on through a combination of colors, that can only happen inasmuch as these combination also elicit characteristic effects in our optic organs. That is why a theory of the harmony of colors must begin from completely different perspectives than that of music, namely, from the concept of complementary color, from the chromatic circle (as an ordering of color according to the concepts of "contiguous" and "complementary") and from the corresponding ideas of symmetry. Goethe's theory of color presents such a theory of harmony. The structures that might render a combination of colors to be of artistic value are, in some respects, more complicated than the structures of music. When we speak of structures of color, we initially refer to the "relation" of colors (they may, in borderline cases, be "complementary" or "contiguous," even though these two concepts in no way exhaust the possible relations.) Then, we have in mind their relationship to the value-binaries light and dark or white and black; and, finally, we also refer to the way colors are spatially arranged which, because of the three-dimensionality of space, is more complicated than the one-dimensional sequence of sounds in time. Spatial distribution corresponds to the rhythm of music but can hardly be determined, let us say, in a painting without reference to the painting's substance. Thus, we have already made connection with the area of specialized symbols which "signify" something quite specific.

Finally, this connection becomes even closer in poetry, where one must go behind the simple meaning of words to find additional spiritual content through structures of language. Here the substance can be carried by the external order, such as verse-structure, rhythm, rhyme, disposition and an internal order that pertains to associations suggested in the meaning of words or in images. At any rate, poetry can hardly be conceived of without the direct meaning of words, while one can barely speak of the specific meaning of an individual symbol in the plastic arts—at least in some of its branches, such as decorative art or architecture.

This raises the question of the relationship between spiritual content that is transmitted in the ordered symbols of art and that which can be expressed in the ordinary means of communication: writing and speech.

It is usually described as follows: The task of art is to move the human soul, to awaken its feelings and to create moods, whereas language (particularly that of science) should mediate knowledge. According to this view, art and science are seen to have completely different tasks and the spiritual content which the artistic order of symbols can transmit differs fundamentally from what can be expressed in the specialized symbols, the spoken and the written word. Even though this view expresses a certain aspect of that relation correctly, we must emphasize nonetheless that the differences cited are slighter than it appears at first, that there likely exists a continuous transfer of the spiritual content of the one kind into that of the other. By contrast, one may indeed speak of an art-work's cognitive value [C49] and compare it to the cognitive value that can be articulated in ordinary or scientific language.

The overly forceful insistence on the difference between scientific and artistic cognition quite likely derives from the incorrect notion that concepts are firmly attached to "real objects," as if words had a completely clear and definite meaning in their relationship to reality and as if an accurate sentence, constructed from those words, could deliver an intended "objective" factual situation to a more or less absolute degree. But we know, after all, that language too only grasps and shapes reality by turning it into ideas, by idealizing it. Language, too, approaches reality with specific mental forms about which we do not know right away which part of reality they can comprehend and shape. The question about "right" or "wrong" may indeed be rigorously posed and settled within an idealization, but not in relation to reality. That is why the last measure available for scientific knowledge as well is only the degree to which that knowledge is able to illuminate reality or, better, how that illumination allows us "to find our way" better. And who could question that the spiritual content of a work of art too illumines reality for us and makes it translucent? One must come to terms with the fact that only through the process of cognition itself can we determine what we are to understand by "cognition."

That is why any genuine philosophy, too, stands on the threshold between science and poetry. The great philosophers were always aware of the "provisory" character of all cognition. They understood or, at least, felt that every formulation of reality in language not only grasps, but also shapes and idealizes it and that this shaping and idealization separates itself from reality precisely to the degree in which the concepts become more and more precise. And for that reason, the ultimate, deepest insights are finally expressed in parables.

Ultimately, cognition is perhaps nothing other than an ordering process—not of something that would have already been accessed by our consciousness or perception but of something that requires this ordering process in order to become the actual content of consciousness or a process we are aware of. The inward illumination that we experience in gaining new knowledge is the unconscious or conscious actualization of that ordering.

(c) Science

Science may be regarded as a specific extension of that domain of reality that can be grasped by the means of communication we call speech and writing. The aim of this widening is, first, to employ those means of communication in a more subtle manner than ordinarily; second, to represent with them an abundance of greater detail in any of reality's domains which would otherwise have eluded observation; and, third, to press forward in this manner to new orders of reality. Since striving for harmonious orders is always the driving force of scientific thinking, science also remains closely related to art. Even where it is a simple case at first of applying scientific methods to practical, external concerns motivated by utility, success comes frequently to the artistic person to whom, even in unimportant details, secret kinds of order reveal themselves, that is, order which cannot be accessed in trivial ways. On the other hand, by being all too active, people frequently fall prey to the danger of grasping the butterfly of understanding with such a rough hand, that they destroy the colorful patterns on its wings before seeing and internalizing them.

We said earlier that reality can be represented by language in two ways; we distinguished between "static" and "dynamic" means of representation. The means of communication which in fact develop already when symbols (or the relations among them) become more specialized and defined, can be made more precise, put into precise and unambiguous contexts, so that finally by this means an increasingly accurate depiction of the domain of reality in question can be achieved. That is how what we called a "static" depiction of reality comes about. Or else the means of communication maintain the degree of variability or ambiguity that is sufficient for communication in day to day life. The manifold interconnections that either came into being among the symbols of language as it developed or were formed on their own in the course of usage are used to tease out ever new dimensions of the object in question, to give it new shape through ever new formulations. In the interplay of formulating, of searching for new links or interpretations, a mental content takes shape that may be taken as the image of the dominant reality under consideration.

The various sciences make use of the static or dynamic form of representation in different ways; correspondingly, artistic elements also impact the sciences at very different places.

Because their subjective matter is much too intricate to permit a language that is absolutely precise and whose interconnections are unambiguously defined, the humanities can almost only resort to the "dynamic" form to represent and order the domains of reality they address. In the humanities, as perhaps similarly in poetry, the objective content of a work can in many cases hardly be separated from the linguistic form of its presentation; great achievement in the humanities presupposes the finished artistic form of representation. The opposite is the case in mathematics. The linguistic form of representation is beside the point; the domain of reality under discussion can be depicted "statically" by the language of formulae which must be exact and free of contradiction and needs no reference whatever to other aspects of reality. The close relationship between mathematics and art derives from the unmediated beauty of the structures articulated by the mathematical statement. Here we can speak of the art of representation only insofar as particular linguistic formulae perhaps make the intended structures manifest in remarkably simple and transparent ways.

To be sure, mathematics and the humanities disciplines share the basic characteristic that their object belongs wholly to the domain of reality that is created only through the existence of symbols. Here the humanities generally accept symbols as they developed among human beings independent of science: as language, as forms of thought, as religious customs, through symbols they maintain the connection to all aspects of human existence. Mathematics, by contrast, dispenses with such connections altogether, demanding instead that its symbols allow for precise, unambiguously determined linkages among themselves. The core substance represented by mathematical symbols surfaces only from those linkages and this fact is at the basis of the often discussed close relationship between mathematics and music. The content that mathematics sets out to order comes into being only through its very activity of ordering whereas in the humanities their content is given from the start as the abundance of the symbols created in the interaction of human beings.

The other sciences whose objects are particular aspects of reality experienced through the senses and amenable to objectification share with the humanities the vast diversity of material, half ordered, half not, from which the new structures of the mind are to be formed. That material may belong to those lower orders of reality we spoke of earlier. It may be ordered in terms of precise concepts that become formed in what we call idealization, concepts that like those of mathematics allow for unambiguous connections. Or it may be elucidated by comparative interpretation from various perspectives thereby letting us see new, previously invisible structures. In any case, the value of a scientific achievement is not measured in relation to its object, that is, according to the meaning the material to be ordered has for human beings, and certainly not according to some arbitrary "practical use," but only according to the beauty and fruitful potency of the articulated structures. In science we are in fact repeatedly amazed by the phenomenon that new structures become linked, as if by themselves, to an existing structure and that the resulting web of structures covers a large area to which the original structure had no reference to at all. This formative power of articulated structures constitutes the real essence of a scientific insight; here the close relationship between science and art once again becomes very clearly apparent.

Here we also discover the answer to a question that is often brought up: Why is it not possible to reproduce great artistic achievements in the style of an earlier epoch or to continue today to gain significant scientific knowledge, let us say, in the area of classical physics or in the Hegelian study of history? [C50] The answer is that the areas that could be ordered in terms of the ideas of those periods have already been ordered in these epochs of intellectual research and the formative power of those earlier ideas has long grasped whatever material was amenable to such a process of ordering. Great scientific achievement comes about only when, in the changing course of time, new material has emerged for human beings to ponder, when the smelting furnace of historical processes releases new, refined material that awaits the germinal nucleus of the crystal that will forever determine its future shape. That decisive ideas in science have often been expressed almost simultaneously and independently by different people is no less natural than that, as the smelting hardens, germinal nuclei often form independently of each other at different places which then often allow crystals to form simultaneously and from different sides.

(d) The Symbols of the Human Communities

As already mentioned, the ability of humans to communicate with one another and the unmediated, instinctual actions of animals that so often amaze us, are mutually exclusive. But, on the other hand, human beings, too, function in biological contexts which decisively determine the course of their lives from the outside, as it were, and from which they actually cannot escape even though they seemingly act freely. In this situation, given that the different layers of reality have to "match," biological reality also has to have meaning in the layer of symbols. That meaning is often established in a particular group of symbols, to which we now address ourselves.

We are dealing here primarily with biological processes affecting a large number of human beings, let us say, for example, the formation of human communities, the so-called political process, and in general those actions that we commonly refer to as the outcome of human drives and passions. When the time comes in the autumn of the year, migratory birds gather in huge numbers as if in response to an inaudible command to begin their communal flight south. Similarly, it may happen that human beings spread out over a wide area all of a sudden become restless and draw into a herd, as if driven by an invisible force, together give in to the intoxication of enthusiasm and wreak some feat of most senseless destruction. Such processes find their expression in the layer of reality of symbols; certain words or concepts, occasionally also mere external marks of identity, become "the symbol of the movement." The words "liberté et égalité" were the movement's distinguishing watchword in France; in Russia it was the red flag and, during the crusades, the cross. Human communities form themselves around such signs and it seems that only the power of symbols decides according to what point of view human communities draw together or separate as "we" and "the others" and often go on to fight one another to death and destruction. The objective difference between two mutually hostile theses can be so small that it fails to make sense in the sphere of logic what a war is being fought about—a clear signal that those theses are merely external signs that represent an event of a different kind in the symbolic layer of reality.

Large human communities always draw together under such a "sign." Since the end of antiquity, the Western world's cultural circle has been held together by the cross; in the Middle Ages, the German empire was united under the imperial crown. A symbol that represents a large community over a prolonged period of time often also becomes the symbol of a specific legal order as the form, transmitted orally or in writing, in which the community's life is to be lived out.

The legal order itself may be regarded as a special sub-domain of the symbolic layer of reality. For in its fundamental elements, a legal order does in fact contain all the characteristic components of that layer of reality. For one, every legal order presupposes the existence of certain specialized symbols: the concepts of justice that alone allow the determination of rights. But then—and this is decisive—a legal order is an "order" after all; it embodies the definite element of this layer of reality, namely the recurrence of like or corresponding things within it. The idea that under like conditions like results should also occur, that it should be possible to predict how people will act in particular situations or how the public will judge or punish an individual's action has terrific persuasive power. Not only does this idea automatically assert itself in every human community, but it can also even form the basis of large communities. If one looks around in history to see what forces hold such communities together, the first thing one finds after the primitive sense of shared race, manifest already in the animal kingdom, is common language. But there are two even stronger forces that can weld together even peoples of different races and languages: a common faith and the strongest yet: a common law. In the process of the emergence and subsequent expansion of the Roman empire, for example, the pre-eminent role was not played by Rome's military might. Rather, the Romans brought law to the peoples they subjugated and the bond that law forges is obviously stronger than all other bonds. Moreover, the expansion of the Anglo-Saxons' domain in today's world—despite the injustice that was associated with that expansion as with all political activity in detail—is based on the establishment of a specific legal code and on the ability, rooted in the foundations of English thought and the English language, to look at human relationships condition from all sides and, hence, to assess them justly and with sober objectivity.

The nearly invincible, community-building power of law is also the reason for the deep hatred of all revolutionary movements for the prevailing law. For, as all hatred grows from powerlessness, here, too, hatred is linked with the awareness that the prevailing law is the real core of the hostile, old community that is to be brought down. It is linked, as well, to the suppressed feeling that this law will prevail in the end in some form or another, all revolutionary forces notwithstanding.

The biological process unfolding in our time among the peoples of the earth and whose uncanny power now manifests itself in all its horror in this Second World War, surely seeks to form larger human communities that extend beyond the national constituencies shaped by common language or race. In this process, the larger communities fighting for their existence gather around very different signs. In Germany and Russia, the higher unity striven for is to be cemented together by means of a common creed although, not unlike in the case of China and Japan, the ideology embedded in that creed leans heavily on the old ideology of the national state. For the community dominated by the Anglo-Saxons, common law and the common affluence it spawns comprise the obvious bond. We do not know how the struggle will end and the result of the war will become known most likely only one or two decades after its conclusion, when it becomes apparent which new communities will go down in history as solid entities and which ones are forgotten for good. Surely in those newly formed entities, there will naturally emerge a new order of law, which alone can be regarded as the external sign of the emergence of a lasting community.

Perhaps one more thing should be added. The individual human being ought never to believe that she or he can actually intervene in the course of world history by means of new ideas or programs. World history is shaped by other and stronger powers and the spirit of the times is not created by human beings. An individual can at most experience the spirit of the time, anticipate its working and give it a certain shape in words. And then, to be sure, those words can be the seeds of crystallization through which the change long in the making suddenly happens as if by a stroke of magic. But even then the individual human being is only a tool and not the driving force of the process. That is why an individual statesman or politician is almost always replaceable—Caesar's work was completed by other after his murder—whereas the great artist is most likely irreplaceable in the same sense. It is true, artists can create only when the time is ripe and when decades or centuries of development have prepared the way for a certain step in the arts. But then, artists create quite independently, their work is truly tied to their person. If, let us say, Beethoven had died thirty years earlier, many compositions would never have been written which, having come into being, have for centuries shown people again and again as they heard them, the way to the real and ultimate things that make them better people and more receptive to the beautiful. A statesman [C51], on the other hand, often hardly needs to be a human being.

It is for this reason, too, that we regard those times when art can flourish as humanity's great and fruitful epochs, whereas the other centuries whose visage was determined by great statesmen, jurists or technologists appear to later generations more as epochs of decline or of preparation.

7. The Creative Forces

Having said all this, now we must finally turn to the highest layer of reality where those parts of the world come into view of which one can only speak in parables. We could begin by telling a parable and talk about that layer of reality which connects us to eternity. But at this point, parables are not yet comprehensible; besides, we first need to look back at the scale of reality's domains which is supposed to reach its apex at this uppermost layer.

The order or map of the domains was meant to replace the rough division of the world into objective and subjective reality. Stretched out between those two poles of object and subject, the order placed the domains at its lowest end where we could objectify the phenomena completely. Then came the domains where things cannot be totally separated from the process of cognition by which we come to determine the facts of the matter. Finally, at the top, there ought to be the layer of reality where such facts are established only in connection with the process of cognition. This way of putting it is open to a twofold misunderstanding: First, when it is seen as a paradox in which cognition is possible only of things that already exist before they become objects of cognition, and, secondly, when it is assumed that the phrase "the fact of the matter" is obviously meant to refer to some sort of subjective illusion that somehow sneaks in on its own as one strives for cognition. In order to elucidate in more detail the connection we have in mind between fact and cognition, we refer once again to the relations in the domains of reality that we spoke of earlier. We begin with an example.

We know that among human beings love exists; one can also often speak of it as of any objective state of affairs. But we have also experienced that the relationship to another human being can be a very tender creation which can be altered by being affected by any word or even any idea. Finally, there can even be human relationships that can only persist in the same form if they never become objects of conscious thought. It is quite obvious in all these cases that any perception of the facts will necessarily have an impact on and change those factual conditions. A brooding person, let us say, who is accustomed to monitoring her or his feelings rigorously at all times will very quickly change such a relationship into something else, while an extraverted, open person can live in such a relationship for a long time without taking note of it even when it has already laid hold of her or his being. Here, too, one's whole condition will change completely once one has become consciously aware of the relationship. Now this epistemological conundrum would not have far-reaching consequences if it were merely a matter of cognition of a particular psychological situation. We know, however, that love transforms all of reality in a much more general and serious way. In our relationship to human beings the world around us changes its face. To be sure, the small part of reality that can be completely objectified does not change here. But wherever things have meaning for us, that meaning is decisively affected by our position toward human beings. The objective brightness and color of things around us, where they can be registered with optical instruments, do not depend on us. But whether the world appears radiant with bright colors or dull and gray is completely determined by our position toward human beings and by the condition of our consciousness. Here this part of reality often has much greater weight for the whole of human life than the objective domain. Happiness and unhappiness are only slightly dependent upon objective, external events. Our happiness requires certain preconditions in our soul and not merely favorable external circumstances. Love makes the soul grow wings, writes Plato in Phaedrus [C52]. Furthermore, this inner predisposition toward the world also determines our thinking and acting, thereby also indirectly reaching over into the objective domain. But, once again, that predisposition depends significantly on the processes of cognition through which we become conscious of it, also, it differs so much from person to person that this part of reality is no longer amenable to objectification. For example, the condition in which we perceive the world as alien, separated from us as if by a veil of fog; can be transformed by a friend's empathetic question whether we are not well. Many more examples could be provided for the same cognitive situation.

We can therefore describe as a first characteristic trait of the layer of reality to be discussed in the following sections the coexistence of two facts: First, reality is dependent to a significant extent on the condition of our soul, that therefore we are ourselves able to transform the world. Second: the effect of this ability to change, nonetheless partially diminishes objectification because, for one, human beings happen to be different and relate differently to the world and, for another, this creative condition of the soul is part of the sea of unconscious soul-processes and cannot be raised up to the surface of consciousness without being altered.

This second point is closely connected to another important aspect: The soul's power to change the world cannot be directed by the human will. Even with the greatest exertion of will power, no one can make the relationship we call love, for example, happen between herself or himself and another human being. On the contrary, an instinctive feeling tells us that the will is an instrument entirely unsuited for handling that part of our soul where the decisive changes of reality take place. Thus, if we say that we can change the world with the powers of the soul, we have to add that we cannot change it according to our will.

However, the human ability to understand is unlimited; therefore, there are also ways for consciousness to exert influence on the soul's creative powers. Religious teachings, for example, at the heart of which is contemplation, contain explicit directions on how people are to comport themselves in order to preserve and strengthen the soul's powers. Probably part of every ethic is basically also a collection of such instruction designed to keep the soul healthy. Only the superficial observer thinks of moral law as an impediment to the individual's life in favor of the community, as a restriction of freedom. For the thoughtful person, it is the collection of ancient insights into how we ought to conduct ourselves in order—as the ancients used to say—"to be happy" or, in Christian terminology: "to find mercy in the eyes of God" or, in terms of what we are addressing here: "to safeguard the soul's creative powers." It will be conceded that these three formulations basically mean the same.

(a) Religion

All religion begins with the religious experience. But one will speak very differently about the content of that experience depending on whether it is encountered from within or from without, so to speak. If the experience concerns us personally, we can speak about its content only in the form of parables. We might say, for example, that we suddenly became aware of the connection to another, higher world, in a way that committed us for the rest of our life, or that in a certain situation God met and spoke to us directly. (I myself would think here first of all of that summer-night in 1920 on the balcony [C53] of the Pappenheim Ruin.) We could also put it like this: The meaning of life suddenly became clear to us and now we know how to distinguish with certainty between what is of value and what is not. "Whoever once circled the flame [C54] always shall follow the flame,"—becoming aware of the other, higher world is something that happens to us quite unexpectedly, from the outside, as it were, so that we can have no doubt whatever that all of a sudden another world stands there before us and claims us. At the same time, this other world strikes us as something we have known for a long time, something that has been familiar to us from the beginning of our life. Just as in returning to a place familiar from childhood, let us say, the smell of the corridor of one's old family home can, as if by magic, make bygone days ever so present again, so the breath of that other world touches us as if it had come into our presence at a time long lost to memory. And whatever the image in which we try to depict what we experienced, the commitment stays with us throughout our entire life and we acknowledge it even when we fall short of it. They who in the course of life would forget that commitment and become indifferent to it lose access to what is most valuable in human life. "But when he loses its gleam, duped by a flash of his own: law of the center he leaves, shattered and driven through worlds."[2] Today this applies to many people, belonging to no religious community, have encountered that other world for the first time, for example, in hearing a Bach fugue or in the brilliance of a scientific discovery. The commitment remains firm for them, too, as does the awareness that since their encounter they know the difference between things that really matter and things that don't.

Looked at from the outside, the religious experience appears as a structural change in the human consciousness and in its unconscious foundation.

---

[2] The two quotations are from the long poem: "The Star of the Covenant" by Stefan George, in: Poems – Rendered into English, New York: Pantheon Books, Inc., 1943, 210. (Transl. note.)

We note that the persons in question have changed their position toward the world and that their words and actions are affected by that change. Observing from the outside, we would hardly think of saying that reality has changed if it had occurred only to a single human being. But then we become aware of a remarkable phenomenon: The same change can take hold of many people. Obviously, what is happening here is similar to love which, if it is genuine, always communicates itself from the lover to the beloved person. Thus, through one person many others also gain access to the other world, as we described it a moment ago in a parable, This access finds expression in symbols which, by definition, separates a community from the rest of humankind. Finally, the content of the religious experience takes a comprehensible form in religious myth, in the parable that creates the language with which the content of religious experiences can be talked about in the first place. If a change of human consciousness has come about in this fashion in large human communities, then it makes sense to speak of a transformation of reality. The fact that there are still people who live by other structures of consciousness elsewhere on this earth is of no great consequence, for within the great religious community everyone understands the symbols of religious myth; for the members of the community they describe real experiences and thus depict a genuine part of reality. The binding character of the religious experience causes the other domains of reality to be included when one interprets things in terms of religious symbols and causes the question of objectifiability to lose importance. What is demanded of truth is no longer that it is objective but binding for everyone.

To help understand this situation, we recall the famous colloquy between Luther and Zwingli about whether the bread of the Eucharist "is" or "signifies" the body of Christ. The way that question was framed signaled a break in the consciousness of the people at that time. That religious colloquy is often interpreted by pointing out that the Christian faith was so firmly rooted in the Middle Ages that it would not have occurred to anyone that it was not the body of Christ. Only the doubts and the sweeping changes of the Reformation period would have given rise to the question whether this was not a matter of only symbolic significance since one could obviously not speak of a material transformation. But it is probably even more accurate to assume that, conversely, it was utterly taken for granted in the Middle Ages to see this as a matter of symbolic significance and not one of material reality (even though the Middle Ages would never have phrased it that way.) For it was only the symbolic significance that was sufficiently important for that age that it could make use of the word "is" or the word "substance";

symbolic significance was the uppermost layer of reality and that is why the bread was "really" Christ's body.

The bonding character of the religious experience also helps us understand that the differences of faith in general create hopeless divisions among people. People of different faiths are divided about essential things. Hence the bitter fierceness of all wars of religion, always waged for the holiest of goods against an unbelieving enemy who in the eyes of the believer is more animal-like than human. For as someone of a different structure of consciousness, the unbeliever is, in fact, almost as alien as an animal whose mere existence threatens one's security.

If one looks at religion and the actions of religious communities in this manner, the power of the human soul to change reality seems to be more calamitous than fortunate and one might be tempted to wish that future generations would refrain increasingly from taking their experiences of a higher world—as it is called in the parable—seriously in this way and from talking about them in symbolic terms.

But that wish could never be fulfilled. For nothing can alter the fact that reality can be changed by our souls. Nor would we want it otherwise, for all of humanity's great spiritual goods ultimately have their origin in this fact. But then again, since religious experiences (in the most general sense) necessarily represent the ultimate standard of value against which all human acting and thinking is measured, human beings will also always form symbols with which they communicate about that standard.

One might object here, pointing out that a large part of humankind, particularly in our day has explicitly renounced any and all connection to religion. But in reality it is the bonds to religions that speak explicitly of God that are being severed, while this makes room for religious bonds of a different kind, those in whose myth, for example, the creative power of the soul is downplayed as much as possible. For part of humanity, turning away from hitherto existing religions is apparently only the preparatory phase in order to enter into new bonds. The development of such peculiar this-worldly religions as National Socialism and Bolshevism signals that here perhaps new and decisive changes in the structure of human consciousness are in the making. For another part of humanity, particularly in the Anglo-Saxon world, a bond of a somewhat different sort has long since taken the place of the earlier religion. This bond is linked back to the experiences of the first great minds of nascent modernity who along with Christian reality rooted in revelation also discovered that other objective reality which began its triumphal march with the emergence of modern natural science. A large part of humankind today looks upon the objectifiable layer of reality as reality per se; it represents the foundation of every standard of value. As in every religion, such unconsciously assumed valuation is based only for a small portion of its believers on the repetition of the experiences of the great minds; the greater number probably take in those experiences only in unclear and vague ways. Still, many people can be affected by the human spirit's impacts on the objective material world; the sight of a giant ship or of the buildings in Manhattan that reach into the clouds can create in us a state of amazement that lets us feel the demonic powers with which we human beings have allied ourselves. Perhaps the persuasive power of the Anglo-Saxon world-view is based in such experiences. But, surely, the question is whether and to what extent that world-view can be compared to the other religions. It does have some features in common with those religions, particularly that here too, the believer will hardly gain access to the world of inward experience of another religion's adherent. Like the other religions, this world-view points us human beings to something that is outside or above us and no longer subject to our will: the eternal laws according to which the objective world functions. But the consequence of the fact that this world-view has no myth that speaks in symbolic form of the soul's creative power, is that at a decisive point it is less significant than the genuine religions. While the real religions again and again direct attention inward and strive to keep the creative domain of the soul as free of injury as possible in spite of all the misfortune in the world, the world-view devoted to the objective leaves the soul defenseless before every monstrosity. The damage done in this process may be even greater because it normally does not enter people's consciousness. That is why it is improbable that this world-view will be able to prevail in the long run, in case the words of Christianity eventually become incomprehensible altogether. Instead, another language will have come into being in which the forces will again be named explicitly that transform the world through our souls.

(b) Illumination

It is not by our will that love and "the other world" come to us. We can perhaps make ourselves receptive to their coming, we can wish that they would come, or have given up every hope for their appearance. Be that as it may, wherever they touch our lives we must simply accept them as a gift, without asking whence they came, as the grace of a higher power that determines our fate and to which we may gratefully submit. If one keeps in mind how this event appears as a change of the structure of the human consciousness when looked at from the outside, one may then say by analogy that such changes are as little subject to our will as, for example, the growth of our body or its healing powers. Yes, exercise and care of the body will keep it strong enough that, when injured, its healing power can intervene undiminished but our will cannot coerce the healing to begin. Something like that holds to an even greater degree for the soul's creative powers which, as a part of the ultimate and innermost forces of all life as such, determine our fate from a higher plain without our will.

This reminds us finally that the creative powers manifest themselves in yet another form. It happens explicitly in this ultimate, most decisive place and only to the one graced human being, namely where we speak of spiritual illumination or of the inspiration of the genius. In all ages, people have seen it [C55] this way whether one speaks like Plato of divine madness, or human beings were taken to be the tool or envoy of God, or whether the ingenious man or woman was revered as a special kind of human being, as was the case in the nineteenth century. It was always acknowledged that without willing it, power flowed into rare individuals to capture in symbols what is intransient, to make God's action known in their time and, in so doing, to intervene for centuries in humankind's fate, fortune or misfortune. Of course, this does not happen without inner preparation which, through years of hard work or difficult human experience, creates the conditions for decisive qualities to manifest themselves. But, of course, this external course of the particular person's life is already part of the task that has obviously been given to this human being from the outset. Also, the task is always openly acknowledged with the maturing of consciousness and made into the plumb-line of life regardless of the sacrifices required. Those with whom this happens are, after all, no longer merely human beings; rather, they are workshops in which the creative powers are visibly at work, creating testimonies that point beyond everything belonging to the human sphere. What takes shape in this highest layer of reality is at one and the same time the most objective and the most subjective: the most objective because the persons in question are aware at every moment of creativity that they are acting at the call of another world that is creating through them, and the most subjective because in each instance, what has been brought into being could be said, written or thought in this particular way only by this one human being.

In connection with the task given to the individual human being, one may ask again what role the human race as a whole plays in the history of the development of the earth or—at least for us—of the world. When we raised this question the first time, we said that the human being likely comes from a central evolutionary series of organisms in which nature managed to preserve a particular line by repeatedly avoiding the specialization of particular achievements and safeguarding the highest degree of adaptability. We mentioned as well that, during growth from the ovum onwards, the individual human being, to a certain degree, retraces the whole range of evolutionary stages and in childhood must also repeat in the course of a few years the whole spiritual development of humanity up to that time. Bearing this in mind, one may ask how long the individual human being remains in that sense at the core of the central evolutionary line, that is to say, participates qua individual in the ascent and progress of evolution and when the individual is eliminated from it, having an ongoing effect on this earth only through descendants or through the traces that were left behind. If one pursues this question, it appears that the whole range of what human development has to offer is wide open for many people up to the end of childhood. The years of transition, let us say from age 13 to 18, as we move into another condition, all life-forces apparently draw together once more in order to let this individual, too, participate in the ultimate and highest that the plan of creation has granted to us human beings at that point in time. But the bud that grows here does not reach full bloom in most cases. When entering into adult life, the decision for many about their task is to reproduce the human race; the tension that had kept the individual life close to the great central line of humanity is loosened and transfers itself to the next generation. The process of development is carried further only in a few. Granted, many do still occasionally become conscious of being woven into that great life-process beyond the bounds of personality. This may happen in times of great passion or when sacrifices are being made for a human community or perhaps the impact of a great work of art. But this, too, gradually fades away like a memory and only few humans remain at the height of the creative forces which continue to labor in the human spirit for something higher. For these few people, it is only the given task that determines their destiny. It is not unusual for the forces to burst the spirit's vessel wherein they are at work and in a catastrophe bring an end to the individual's mental or physical existence. One thinks of Hölderlin or Hugo Wolf [C56]. In other cases, the person's bodily strength simply seems insufficient in the long run to withstand their excess of spiritual energy. Mozart and Schubert come to mind. In the end, very few are up to bearing the burden of such a task for their entire life. In the course of the years, the work of such creative artists jettisons everything coincidental and personal, everything that bound them to an earlier phase of development long overcome. And so, at the close of such a life, there are those utterly pure works of the spirit, detached from everything earthly, such as the conclusion of "Faust" or the measures:

According to Plato, love is the longing of human beings for immortality and that the holy terror in face of beauty is at the same time the terror in face of the infinitude that suddenly floods our consciousness in such moments. Perhaps one may also say that not only in love but in all those moments when the "other world" encounters us, a feeling arises in our consciousness for that infinite life-process in which we participate for a short while and which is realized within us and beyond our earthly existence.

(c) The Great Parable

Everything that has been said here can also be framed as a discussion of the perennial question about the existence of God. As it has sought to answer that question, human thought has taken many steps, each of which is essential in order to reach the next.

The first thing we could say was simply: "I believe in God, the Father, the almighty creator of heaven and earth." The next step—at least for our contemporary consciousness—is doubt. There is no god; there is only an impersonal law that directs the fate of the world according to cause and effect. Therefore, it is self-deception to want to talk about a personal god to whom we can turn. What we discover and take to be the order and harmony of the world is merely the working of eternal laws or of the ordering power of our mind.

The next step, would perhaps be Voltaire's frivolous formula: If there were no god, someone would have to invent one [C57]. That is, faith in a personal god is at least a useful, allowable self-deception, a self-deception that makes for the harmony of our soul.

But all such formulations only derive from a first and preparatory reflection about what we are talking of now. For after we have pursued every one of these sequences of ideas, we become aware that we do not know at all what exactly that word "god" and, especially, what the words "there is" mean. The words "there is" belong to the human vocabulary and refer to the reality that is reflected in the human soul. We cannot speak about another reality. But if the words "there is" cannot signify another reality, then their meaning changes just as reality changes with our faith. One can speak about the ultimate ground of reality only in terms of parable and when people say parabolically "I believe in God, the Father," then this god really directs the destinies of human beings like a father through their faith. This faith is no self-deception but simply the conscious acceptance of the ever irresolvable tension in reality which surely is "objective" and runs its course independently of us humans. And yet again, it is also only the content of our soul, and it is transformed from within our soul. But the same state of affairs can therefore also lead people in the opposite direction. For example, when a significant number of people today insist that the word "is" should really be applied only to the objectifiable dimension of reality, then the world also "really" only moves according to cause and effect, without any higher "meaning." Thus, it seems in the end to be simply a matter of people's belief whether a gracious father directs the world's destinies or whether the law of cause and effect mercilessly determines all human fates.

But even with this insight we are only at the beginning of this endless problem. It may well be true that all the great parables: the personal god, the resurrection of the dead, the transmigration of souls, are reality as long as people have the strength to believe them. But then, ought we not to renounce so subjective a reality, one that throughout the centuries has seemed to be unstable, and restrict ourselves to that domain of reality that can be objectified and will surely outlast even millennia? That seems to be the stance many people seek to uphold today. But it is a position also based on an illusion: the assumption that it is possible to avoid changing our world with the force of our soul. But the very declaration of the belief that the objectifiable layer of reality is the "actual" reality changes or determines reality in a similar way as does any other belief; hence, we are just as dependent as before on reality being conditioned by subjectivity.

It would seem, therefore, that reality is in a sense handed over by means of people's belief to their arbitrary will and that humanity's great religious wars, for example, (presumably including the current war) can be seen as nothing else than genuine decisions about how reality is to be shaped. In face of this gruesome possibility, it is consoling to the human mind to realize that faith itself is not dependent on our caprice but that it comes to us without our doing and that, seeing how it was laid upon us by our fate or our age, we need to accept faith as a gift or as our undoing. Of course, even then we may also either give ourselves over to a faith or resist it, take our place either within a human community or outside it. It surely is fortunate that there seems to be some room left here for one's own responsibility or moral consciousness to intervene. But in the larger scope of life, it is a higher power that decides about the belief of human communities.

In light of these reflections, it appears that today we may no longer say with the assuredness of children: "I believe in God, the Father, the almighty creator of heaven and earth." And yet, we may with full confidence place ourselves into the hands of the higher power who, during our lifetime and in the course of the centuries, determines our faith and therewith our world and our fate. As Goethe once called a time of great passion [C58] a gift comparable to an especially good vintage year, perhaps humankind may likewise gratefully receive the centuries of new belief as a gift despite all calamity, fully confident that the present episode of its history also will in the end bear good fruit and serve a more noble development. That is why we as human beings have the obligation and privilege to believe in the meaning of life, even when we understand that that the word "meaning" is only a word of our human language which we can hardly imbue with any other sense than that it justifies our trust. But then trust is perhaps our ultimate possession.

The question of the existence of God has long ceased to be a scientific question; it is a question of what we are to do. But that is always very simple even as times change. We are to be active members of human society, helping others and being diligent about it. Thus, in the symbols of the community the foundation of the world remains alive and fruitful to which we feel we belong as harmonious members. And this blossoming in the world, which is at the same time "God's world," finally also remains the greatest happiness the world is able to offer: namely the consciousness of home.

III

We are living in times that, in their unrest and calamity, are threatening the values which had up to now appeared secure to us; if we take the political turmoil as an indicator of the movements in the foundations of thinking, then the catastrophe of these decades leads us to conclude that the bulk of human thinking is shifting and thereby dislodging the foundations. What the world will look like when this process is finished, we do not know. But one is probably reading the signs of the times correctly when one assumes that those domains of reality that we ourselves are unconsciously shaping will regain importance against the domains of the objectifiable; for now, however, the sinister demons of those domains appear to be playing the leading role. Perhaps this shift will go so far as the one at the beginning of our time calendar, so that the connection to the past can only be maintained by very small and isolated groups of people. But perhaps the same thing will not repeat itself entirely, perhaps the essential element of cognition from the past will remain and the shift will come to a halt at the place where the interrelatedness and juxtaposition of the various layers of reality do not seem contradictory but are tolerated as fruitful tension. Individuals cannot do anything in this respect but prepare themselves internally for the changes that are happening without their doing.

Earlier generations were able to build on the work of their ancestors. For our times the goal is of necessity a more modest one, since the old spiritual values have been molten down. Initially, we have no choice but to return to that which is simple; we must conscientiously attend to the duties and tasks which life itself places upon us without much questioning the whence and whither; we must pass on to the next generation that which we still deem beautiful, rebuild what was destroyed and put our trust in other people across the din of passions. And then we have to wait what will happen; the new does not have to be visible right away, we will accept this as it is—reality will change by itself without any action on our part.

When we ponder the times to come, the greatest threat seems to come from the confusion between good and evil. Especially in an era where the ties to established religion are loosening, the danger that demons are talking hold of the rule of the gods is greater than ever; and demons have always allied themselves with that glittering phantom that leads people astray at all times: political power.

So that our vision is truly focused, we have to remind ourselves above all that political power has always been founded on criminal behavior. This does not change or is made better when political power, imposed as order onto a great human community, eventually also brings about things beneficial. In the process of expanding their power, people are, however, always trying to force by brutal means those into becoming members who do not integrate themselves into the community. The banal slogan: "If you don't want to be my brother [C59], I'll crack your skull!" applies still today for each and every one of the large political blocks. That is why we need to be suspicious here; yes, there will always be political power and countless hordes of people will suffer and die in the struggle for political power. But that makes no difference, nothing about the value of the cause people are dying for is being decided by it. Perhaps the new meaning that this world is to be given will go on growing for quite some time far away from political power and unnoticed by it, until, one day it connects large communities seemingly out of the blue.

We probably also will have to content ourselves with the fact that there are and will continue to be large masses of human beings whom–speaking metaphorically-the Good Lord can no longer reach. That cannot be alleviated by compensating these masses with material goods. We are not talking here at all about those who really are poor or who cannot think; we are talking about the very ones for whom the world bears a gray and rigid countenance. If they are within a large, cohesive community, led by the smaller group who see things differently, or by its young people, they are still somehow part of the meaning that connects everybody. But in a time when this meaning has become murky and has to be found anew, they are in a hopeless situation where even the care of others does not make it any less desolate. But that too does not matter. Perhaps nobody can save these masses from the fate of having to fight bitterly on whatever side for political power. But what is crucial now is that the few for whom the world still radiates with light stick together and recognize one another across what the others put in their way. For only to them will the meaning be unlocked that is to be given anew to the world.

The powerful figure who assumes the right to destroy the enemy and who throws resisters in jail is not important [C60]; it is instead the jail guard who, despite orders to the contrary, cannot refrain from slipping a piece of bread to the prisoners now and then. We need to remind ourselves again and again that it is more important to act humanely towards the other than to fulfill any professional obligations or national obligations or political obligations. Even the din of great ideals at its loudest must not confuse or hinder us to hear the one, soft tone on which everything hinges. It has been said so often that weakness will perish, that only the strong will prevail in the struggle of life. That may well be true. But what is the strong? In music, the loudest passages are often not those when the full orchestra fills the whole hall with sound but the bars when a single violin softly sings a melody. Therefore, those who still know the white rose or who can still discern the sound of the silver string must now join together.

Perhaps, in the future configuration of the world, science will play an even more important role than before. Not so much because science is associated with the preconditions of political power, but because science is the place where people of our time stand up to meet truth. While in political life a constant change of values and the battle of mendacious ideals against different mendacious ideals cannot be altogether avoided, in science we enter a realm where in the end what we say is either true or false, where there still exists a higher power that makes a final decision independent from our wishes and, therefore, creates validity. Most important are therefore particularly those areas of pure science in which practical application is irrelevant but rather where pure thought is searching out the hidden harmonies of the world. That innermost realm, where science and art are almost indistinguishable, is perhaps the only place where humankind today meets truth, totally pure and no longer veiled by human ideology and desire. True, the great masses of people nowadays have just as little access to this realm as people in earlier times had to the inner sanctum of the temple. But the masses are quite content to know that some people are entering that realm and that there deception is *not* possible because there the Good Lord decides.

As long as this central realm of science remains untouched, the danger conjured up by the fact that we control the forces of nature to a much larger extent than in earlier times, may not be all that great. These forces can in their effect be directed toward good as long as their order derives from a center that is not set by us but by a higher power. Only when the regulating center is missing will the forces lead into chaos. Stephan George's poem also applies to humankind as a whole: "The one whoever circled the flame." In a way, an order is replicated now on a large scale that used to exist in earlier times among primitive tribes. Although against his own will, the scientist has become the people's magician to whom the forces of nature are obedient. But his power can turn to good only when he is at the same time a priest and acting merely as mandated by the deity or by fate.

Hence, it is also probably no coincidence that it is precisely in science where the transformation of reality has manifested itself more clearly than in any other arena. Here fate itself shows us the way the world is treading, no longer via the detour of humanity's experiences and suffering imposed on it in its history but in the direct attempt to find the truth. That is why recognizing the limitations of objectification is probably more than just another new scientific experience following after many others; rather, it means that we have to address *that* side of reality which does not let us separate what we recognize from the process by which we have come to recognize it. But then that insight itself is just another building block, not unlike the idea of the existence of atoms earlier, or of the sun's placement in the planetary system. Several hundreds of such building blocks only make it possible to lay the foundation to an understanding of reality. In the end, all cognition rests on experience and nothing can foreshorten the journey of thought as it makes its accommodations over the centuries. Often it takes a century of experience to produce a *single* new, decisive thought. Consequently, in order to answer the question what reality really is [C61], one can hardly reply with anything other than the old fairy tale: "How long does eternity last?"

> At the end of the world, there is a mountain, all made of diamond, and every hundred years a small bird flies there and sharpens its beak; and when the whole mountain is worn down, only one second of eternity will have passed.

Commentary on Werner Heisenberg's "Reality and Its Order"

Ernst Peter Fischer

Preamble to the Commentary

As Helmut Rechenberg (1937–2016) explains in his Introduction, the editors of Heisenberg's Complete Works published the manuscript under the rubric non-scientific, i.e. commonly accessible works. Titled '*Reality and its Order*' [*original German title "Ordnung der Wirklichkeit*], it was written in the war years 1941/42, was quite personal in nature, and thus most likely not intended for publication. Therefore no reason existed for Heisenberg to list any sources when alluding to or citing lyrical poems, when he made use of poetic passages in fairy tales, followed various completed and expanding philosophical trains of thought, or mentioned some musical bars and themes, or the scientific ideas and historic developments of his time. These commentaries attempt to remedy this, as much as possible, while also showing the historical context he himself was living in—thus revealing the cultural bedrock which allowed for his thinking and work.

As the writer of the commentary, I want to first address two aspects which seem relevant for understanding Heisenberg's text. For one, to imagine the dramatic life situation the author was in. His homeland was, at the time

---

The text passages that are referred to in this commentary are identified by page number. Where a particular phrase is extracted for comment, it is additionally labeled by the comment number [C1] etc., as given in the main text.

of his writing, involved in a terrible war with no end in sight, and where one did not know which would be more devastating, a German defeat or a national socialist victory. Heisenberg's emotional state in this situation may probably be best characterized by the term Melancholia, although the Romantic period and its art, to which Heisenberg is often referring, would have used the term 'Weltschmerz' here. Heisenberg—at the time a father of four children, responsible for his steadily growing family—must have been pained and saddened to watch, helplessly, as "the political power… by criminals" (p. 119) is misused to ruin his beloved homeland and its grand culture. Furthermore, he must have suffered because, of all things, his fundamental science, theoretical physics, was slated to be a decisive factor in the ongoing war, and that he himself was solicited and ordered to participate. One cannot but notice that Heisenberg, as a way of life, does evade the external threat by turning inward, for instance, by giving lectures on Goethe's theory of colors, as Rechenberg writes in his preface. Accordingly, he also puts his thoughts of an "Reality and its Order" on paper and thereby, in his middle years, comes to a sort of assessment of his life, so as to reflect on his personal fate, but he avoids turning his attention toward the future.

The second aspect, worth mentioning before the commentaries, is the term 'reality' itself. Heisenberg's essay, early on, addresses this topic, which is both profound and all encompassing–"what, actually, is reality" (p. 19), a question he thinks must always be "explored and answered anew". While many people, in their daily lives, like to speak of *the* or *a reality* and assume they can use the word without any problem, the physicists in the early decades of the 20th century had come to acknowledge that one has to be careful, at least as far as the reality of atoms was concerned. There were properties in the innermost world which must be understood as only "having the possibility of being" and not really and actually existing. In quantum physics, to which Heisenberg contributed so incredibly much, the thought of the philosopher Aristotle is revived somewhat, namely that in addition to the physical reality—the res extensa as Rene Descartes would have said—there is a realm of conceivable possibilities—a kind of res potentia—and that the elementary particles of matter are found rather under this umbrella than in actual reality itself. The philosophically relevant point being that Quantum Mechanics, no longer a classical science of reality, owing to Heisenberg and other physicists, allows one now, for the first time, to formulate a theory of becoming. It is only slightly orienting itself on Plato, so cannot be termed platonic, but much rather on Aristotle. The (stable) *order* of reality is more of a (dynamic) *structuring* of reality, which is constantly in flux and formed anew by creative people like Heisenberg, and as he emphasizes, during the 20th century, essentially through the efforts of the natural sciences (p. 20). With this formative thought, Heisenberg finds solace amid the pain during wartime and the political confines of his endangered existence. This is to say that, based on his very experience with physics, (p. 28), he sees the "ability of humans: to understand, to deal productively with reality, definitely without limits". Amidst all the threat and despite all the melancholy, Heisenberg, deep in his heart, carries the certainty within, "that, measured in a human time frame, it will always continue to go on, life, music, science –although we ourselves can only participate for a brief time", as he writes, not here in "Reality and its Order", but rather at the end of his autobiography "Physics and Beyond" [1]. To Heisenberg's mind, humans always retain "the multitude of possibilities" which the world makes available to them, and as he asserts in this essay (e.g., pp. 22, 56). The point is to reach for this abundance again and again and to use it to order reality anew all the time.

The Comments

[C1] "An infinitely exciting game" – p. 19

In 1938/39, the book, titled "Homo ludens", by the cultural historian Johan Huizinga, was published, which deals with the "The Play Element of Culture" [2]. Huizinga is linking to Friedrich Schiller (1759–1805) who expounds in his "On the Aesthetic Education of Man" [3], from 1793, that it is play which brings out the totality of human capabilities. When the young Heisenberg first studied physics, his science was exactly that— an unending exciting game, which ultimately dealt with totality. One may add that Huizinga' s book was republished later on in "Rowohlt's Deutsche Enzyklopädie", together with Heisenberg's treatise on "The Physicist's Conception of Nature" [4].

[C2] "Three seemingly completely independent, yet inherently related events" – p. 20

The journey of Columbus (1451–1506) represents for Heisenberg the prototype of an undertaking where "actual new ground" is gained, as he states in his autobiography "Physics and Beyond" [1]. He views the great accomplishment of Columbus as the "Decision to leave behind every known land and sail west to the point of no return, given that their provisions would be depleted". A scientist must, like Columbus, summon up the courage "in a sense [to] jump into the void", as Heisenberg himself would do in 1925 in Heligoland, when he was the first to find access to the atom, formulating a first version of Quantum Mechanics. (More on this in my biography "Werner Heisenberg- a Wanderer Between two Worlds" [5]).

The feud about Holy Communion between Martin Luther (1483–1546) and Huldrych Zwingli (1484–1531) took place in Marburg. For Luther, Jesus was physically present during communion, while Zwingli viewed Holy Communion simply as a symbolic act of commemoration. Heisenberg, on p. 111, is returning to this conflict that was not resolved in the 16th century, leading to a split not only in the Roman church, but also in the Reformation movement.

The name Copernicus (1473–1543) remains to this day connected to the notion of a revolution, with physicists primarily thinking of the idea of a heliocentric world view, where the earth is revolving around the sun. Philosophers are thinking instead of the second rotation of the earth around itself, so that the observation of movement of stars in the sky is based on the rotation of the human observer on his planet. Immanuel Kant (1724–1804) in his "Critique of Pure Reason" (1791) termed this the Copernican turning point of metaphysics.

[C3] "The turbulent and fruitful years ..."-p. 20

"The turbulent and fruitful years after World War I" connote the "Golden Twenties" or the Roaring Twenties, during which, in Germany, a Weimar Republic was tentatively established and physicists had a breakthrough to the modern version of their science, namely quantum mechanics. The decisive years in this respect were the years between 1923 and 1927. They contain the "Revolution in Physics" as it was termed in the book by Ernst Zimmer [6] For Heisenberg, during these years of his youth, the world was filled with the sound of the silver string, which he mentions further on, although, in the 1940's, it has long since been torn apart.

[C4] "The sound of that silver string", p. 21

"The sound of that silver string of which Gottfried Keller (1819–1890) has sung", is often mentioned by Heisenberg (e.g. other times on pp. 90, 120). The Swiss poet and writer, Keller, brings the silver string to sing in his poem from 1845 "Jugendgedenken", ["Thoughts of Youth"] with its first verse intoning:

I want to see myself reflected in those days,
That flew away like the wind in the crowns of linden trees,
Where the silver string, when struck,
Clearly, yet trembling, gave the first sound,
Which my whole life long
Even today resounded,
Though the string was long since ripped.

When the "naughty lyricist", as Keller referred to himself, heard the sound of the silver string, he was taking part in the marches of the so-called Freischaren (armed insurgents), which faltered, however, as they were deemed anti-clerical insurrections in Switzerland. Keller's biographer Adolf Muschg (*1943) considers "Thoughts of Youth" the author's most beautiful poem—his fiction features a bourgeois realism. Heisenberg heard the "first sound" of the silver string in the 1920's, and he had to experience, as his life went on, how the source of the sound was destroyed after 1933, and no "Morning star of good intention" was "shining gaily "any longer, as it says in the sixth and last verse of Gottfried Keller's poem.

[C5] "People and things ....remain mute and look rigid"- p. 21

In 1899, Rainer Maria Rilke (1875–1926) published a poem, with the title "I am so Afraid" where he is fearful of "the word of humans" and advises them to, "stay away", "I love to hear the singing of the things, /you touch them: they are stiff and mute, /you are killing all the things of mine." Like the poet, Heisenberg is reacting in fear and counters it on the next page with the "abundance of possibilities" humans can grasp and which will always be open to them.

[C6] "Augustine's Confessions" – p. 22

The many available editions of "Confessions"—"Confessiones" in Latin—by the church elder Augustinus (353–430) were written about 400 after Christ and speak, among other things, of his turning away from the world and his turning towards an ascetic-celibate life. In his book, Augustinus is also asking the question: what is time, wanting to know, for instance, whether there was a time before creation and what God was doing during this time. In his deliberations on time and eternity, the church elder looks to the Greek philosopher, Plato (428–348 B.C.), and is allowing access into his own emotional life.

[C7] "Uexküll's studies on the animal world."- p. 22

This refers to Jakob Johann von Uexküll (1864–1944), a philosopher and zoologist who introduced the term environment into biology, and who is considered a founder of ecology. In 1909, his book on "Environment and the Inner World of Animals" [7] was published; the second edition appeared in 1921. In 1920 Uexküll authored a book on "Theoretical Biology" which, however, never reached the status of Theoretical Physics.

[C8] "But in the perception of the great age of natural science"- p. 23

Heisenberg addresses the epoch of classical physics which ended in 1900, as the quantum of action appeared. During that time, mechanics, electrodynamics and thermodynamics—mostly a kinetic theory of gas—, originated, and during the 19th century, Hermann von Helmholtz (1821–1894) [8] dominated the natural sciences as the "Chancellor of the Reich in physics". Like his colleagues, he dreamed of a complete theory. Although he also investigated observations of colors and sensations of sound, he assumed to be able to relate everything, ultimately, to physics, and to offer a complete causal order of reality. When Heisenberg entered the world and came to physics, this dream had long since evaporated.

[C9] The "element of indeterminateness"- p. 24

Heisenberg's fame is mostly due to the "uncertainty relations", which he established and published in 1927. These relations state that properties of atomic objects—for instance their position or their momentum—remain uncertain, as long as they are not determined—literally: put in place—by a measurement or observation, and that one cannot know both the position and the momentum exactly at the same time.

An electron possesses no position, unless one is found via an experiment, and thereby becomes established. If in the text there is mention of the "indeterminate nature of language"—which is reiterated on the following pages—then there now becomes apparent a commonality in the building (ordering) of physical and symbolic reality. This permits the beautiful thought that man can have freedom (because the whole world is not so much predetermined but rather undetermined).

[C10] "How does a child learn language?" p. 24

Commentary on Werner Heisenberg's "Reality and Its Order"

During the 1920's the Swiss psychologist Jean Piaget (1896–1980) had, among other things, begun to study the cognitive development of children, and attempted to also arrive at a "genetic theory of cognition". He was interested in the acquisition of language in children in this context. The French original edition of his seminal work on "Theory of Cognitive Development" [9] appeared in 1926.

[C11] "Hegel's dialectics"- p. 27

The "Hegel Dialectic" refers to the philosopher Georg Wilhelm Friedrich Hegel (1777–1831) who himself had deemed his predecessor from antiquity, Heraclitus (ca. 500 B.C.), to have been of the Dialectical School. Heraclitus perceived in the constantly changing world a permanent battle of contrasts, accompanied by dualities or polarities, which stood in contradiction to each other. Ever since Hegel, one can understand dialectics as the art of conversation or debate, in which one thesis is contrasted with an anti-thesis and out of this initial situation a new understanding, a synthesis is developed, ending the activity of thinking. In the philosophy of quantum mechanics, as developed by Niels Bohr (1885–1962) together with Heisenberg, known as the 'Copenhagen Interpretation', the tension between thesis and anti-thesis remains active and thus also the demonstration and order of reality, as Heisenberg writes. Bohr terms this state complementarity; it denotes an equally valid counter position of non-compatible thoughts, which can be interpreted now as qualitative dialectics. Heisenberg in "Reality and its Order" keeps referring, time and again, to complementary aspects or positions.

[C12] "About the ultimate things we cannot speak"-p. 29

In the year 1921, the Austrian philosopher, Ludwig Wittgenstein (1889–1951), presented his famous treatise with the title "Tractus logico - philosophicus", the last line of which says "That whereof one cannot speak one must be silent about." Heisenberg is referring to him, although his trust in language surpasses Wittgenstein's. Heisenberg once wrote "That whereof one cannot speak, one must reach an understanding, one must have a dialog, and it is the task of the scientist to undertake it, in order to prepare the way to that world which he is equipped to find, and which to know is his privilege". At the end of this "Reality and its Order" he embodies a position one might characterize the following way, "That which one cannot talk about, one must tell a story about, as, for instance, in a fairytale". As far as the limitations of words are concerned, to which Wittgenstein is alluding, Heisenberg, being a stellar and proficient musician, knows the way art offers a way to express oneself in sounds. Heisenberg goes there himself, copying the musical bars on page 115/116 and discussing them.

They are from a piano sonata by Ludwig van Beethoven (1770–1827) to whom musical experts ascribe having erased the limits of verbal expression in his musical compositions. Heisenberg certainly felt this, when he played Beethoven's music or listened to it, be it the violin concerto or the serenade for flute, viola and violin in D-major by the young composer. Through music and owing to it, Heisenberg regains time and again a growing trust in a central order. It gives him the inner strength to endure the malevolent years of the Third Reich.

[C13] "The pure ideas… of the soul", p. 29

The "pure ideas of his soul"—to which Heisenberg is referring here come from the writings of the French philosopher Nicolas Malebranche (1638–1715). He attempted to reconcile science and church, for instance in his reflection on the human soul. He collects these contemplative thoughts in his book "The Search after Truth" [10], published in Paris in 1675, and in 1914 in a German translation. Heisenberg, even while in school, had discussions about Malebranche, in order to understand "how the human soul arrives at its conceptions" and how the ordering tendencies in the world "are also affecting the formation of the human soul and manifest in it. ("Physics and Beyond" (Ref. [1], pp. 13–14)). Heisenberg's interest in the soul is probably based on Plato's writings (428/7-348-7 B.C.), where the Greek philosopher endows the body of man and the matter of the cosmos with souls—one body soul and one world soul.

[C14] "The Pythagoreans' studies" p. 30

Heisenberg again and again comes back to the Pythagorean discovery "that oscillating strings with equal tension will sound harmonically if their lengths are in a simple numerical ratio to each other"—here, e.g. p. 98, and later in life in his talk on "The Importance of Beauty in the Exact Sciences" presented on July 9th, 1970. The name Pythagoreans denotes the disciples of Pythagoras (570–510 B.C.) who declared "everything is number", and understood mathematical structures, i.e. rational number states as the origin of harmony and beauty. Heisenberg followed this idea unconditionally.

[C15] "Plato's ideas about symmetrical bodies" p. 30

Platonic bodies, named after the Greek philosopher Plato are polyhedrons with the largest possible symmetry. Only five different bodies exist: Tetrahedron (four-sided), hexahedron (six-sided), octahedron (eight-sided), dodecahedron (twelve-sided), and Icosahedron (twenty-sided). Plato describes them extensively in his dialogue "Timaios", where he correlates these bodies with the four elements of Antiquity—fire, earth, water and air—and the fifth, following a postulate by Aristotle, with an ether. This idea continues today as "quinta essentia", which is quint-essence. The underlying idea is about mathematical structures, something Heisenberg would often attempt to incorporate into aesthetic speech. "He compared the number theory with Bach's 'The art of the Fugue', both of which he admired with a kind of mystical shudder" as his complicated friend, Carl Friedrich von Weizsäcker (1912–2006), wrote. ("Zeit und Wissen" [11]). Bach's fugues are mentioned by Heisenberg on p. 30.

[C16] "Beauty of a melodic line" p. 30

The secondary theme in the first movement of the D-major violin concerto by Beethoven (picture 1) is played first by the woodwinds, gets augmented by the kettle drum, and taken over by the violin, and finally played in d-minor (bar 43–58). It resembles the Silesian folksong "O joy above joy".

[C17] "that indubitable statements are ..." p. 31

Immanuel Kant (1724–1804) differentiated analytical judgements from synthetic ones—"The sphere is round" is analytical, "the sphere is red" is synthetic. Beyond that, he differentiated between judgements a priori (before any experience) and judgements a posteriori (after an experience). In the 20th century, the behavioral scientist, Konrad Lorenz (1903–1989), who, for a while, held Kant's chair at the University of Königsberg (now Kaliningrad), looked at "Kant's theory of the a-priory in light of current biology" [12]. He made a distinction between something that is a priori for an individual, must have been learned over the course of evolution, and thus be considered a posteriori for the species. Heisenberg must have known this text and comprehended its significance during the war. Lorenz, already in 1941, develops the fundamental aspects of his Evolutionary Theory of Knowledge, which, however, would only find its followers and elaboration in the late 1960's, for instance in the book by Konrad Lorenz "Behind the Mirror" [13]. One could also mention that Charles Darwin (1809–1820) in a notebook points to an idea of Plato that belongs to this topic. Plato writes in his dialogue, "Timaios", "that our vital ideas stem from the preexistence of our soul", which Darwin quotes, adding: "Read: ape as preexistence". Kant and Lorenz are mentioned on p. 96 again. And one more thing: The physicist Ludwig Boltzmann (1844–1906), made first mention, in 1905, in "Popular Writings" (Braunschweig 1979) of the idea that one can view "thought principles" as" inherited thought habits" since they "are innate to the individual through the many thousands of years of experience within the genus", and therefore can be termed "a priori".

[C18] "Archimedes' laws of levers" p. 33

Archimedes (287–212 B.C.) is among the significant mathematicians of antiquity. His law of levers—force times force-distance equals weight times weight-distance—was formulated when he pondered the "equilibrium of level planes". He supposedly said, "Give me a fixed point and I will take the world off its hinges". In atomic physics, the physicists in Heisenberg's century almost succeeded.

[C19] The meaning of life ..." p. 34

Heisenberg repeats this sentence, ascribed to Niels Bohr, in the 11th chapter of his autobiography "Physics and Beyond" [1], as he writes about the "discussions on language" with the eminent Dane. Heisenberg adds Bohr's further comment "this is how bottomless all this striving for knowledge is". One must add, for one, that Bohr rarely expressed himself as clearly as Heisenberg can, when he supposedly is quoting him. The best wisdoms of Bohr we owe to Heisenberg's recounting. In the chapter mentioned above, he shows us Bohr, doing the dishes in an alpine hut, after a joint skiing trip, and then explain what one can learn from such cleaning in the following words: "Washing dishes is just like using language. We have dirty dishwater and dirty kitchen towels, and yet one can get the plates and glasses clean, ultimately. The same in language - we have unclear terms and a logic which is limited, to an unknown degree in terms of its targeted use, and yet one succeeds and reaches clarity with respect to our understanding of nature".

Secondly, one must note that Heisenberg expounded extensively on the topic in "Language and Reality in Modern Physics". At first in the book "Physics and Philosophy" [14], which has a chapter on language and reality, and then, more extensively in an essay with the above title.

[C20] "A paragraph from the supplements to the theory of colors", p. 34

"Goethe has accompanied him throughout his life", as Elisabeth Heisenberg (1914–1998) said about her husband, and as Helmut Rechenberg (1937–2016) mentions in his preface. Rechenberg also makes clear that Heisenberg revered Goethe (1749–1832) from his early youth; he had the complete works of the "great man" in his library, and he knew by heart "Faust" and many of Goethe's poems. The concept of a multiple-layer model of reality—which Heisenberg borrows from the Additions to Goethe's Theory of Colors (p. 788, vol. 25 of the Complete Edition of Goethe's works, published in 1989 in Frankfurt)—was the topic of a lecture Heisenberg gave in Budapest in 1941, titled "Goethe's and Newton's Theory of Colors". Heisenberg thereby assumes a "division of the world … into several interrelated sections which are segregated from each other, depending on the questions we ask about nature, as well as the interference we allow during the investigation." One must note that, in 1940, the first edition of the book, titled "The Construction of the Material World" appeared, where the philosopher Nicolai Hartmann (1882–1950) develops a multi-layer model of reality, encompassing four stages—the inorganic, the organic, the spiritual, and the mental—which found great agreement among the readers, interested in natural sciences. Among them Konrad Lorenz who writes, in his already cited work "Behind the Mirror", that the sequence of Hartmann's categories of existence is consistent … with the sequence of their origins in terms of natural history. The inorganic on earth has existed long before the organic; and during the course of evolution of the species, there appeared only rather late central nervous systems to which one … could attribute a "soul". The spiritual then entered only in the very latest phase of creation.

[C21] "the ways substance behaves" p. 35

Here the term substance appears for the first time which will become more prevalent in the next passage. Heisenberg surely does not refer to a chemical substance or the like with the word substance, but much rather to something independently existing, as is customary in philosophy.

[C22] "For us, 'there is' simply only the world…" p. 39

In 1919, the philosopher Martin Heidegger (1889–1976), dedicated his first university lecture to the question, "what does it mean 'it exists'?" To him, the fundamental question at the time was 'does anything exist?', according to his view that "only for humans there 'exists' a world". Therefore, because of this one can replace the term "human" with the term "existence", something Heidegger then is doing in his writing (cited from Wolfram Eilenberger [15])

[C23] "Newtonian physics begins with the concept of substance" p. 39

A peculiar observation, for if one studies his mechanics, one will read about mass-points, bodies, forces, and acceleration, while the word substance is nowhere to be found. Presumably, Heisenberg, by pointing to substance, wants to indicate that the mechanics does not deal with fields (p. 42) or imaginary objects that he himself has been investigating. Substance means something basic and is used in classical metaphysics, such as by Newton's contemporary, G. W. Leibniz, (1646–1716). Newton rather is referring to a mass which he ascribes to particles. It endows bodies, once they are pushed, with inertia which maintains their motion. Newton distinguishes mass from weight, and both properties are inherent in material bodies, i.e. matter, which has been a difficult concept and, ever since Aristotle, has had a complicated history. The term substance arrived in physics from metaphysics over the course of the 18th century, when one asked "what is heat, the effect of which can be sensed easily". The Scotsman, Joseph Black (1728–1799), suggested at the time to explain heat as a heat-substance '*caloricum*' which still lives on today as calorie. Heisenberg writes of it in a separate section starting on p. 52.

[C24] "A perpetuum mobile" p. 42

Perpetual motion refers to a device that is incessantly moving and thereby doing work. Regrettably, such a device cannot be constructed, due to the physics law of the conservation of energy. Nevertheless, people attempted, for a while, to imitate on earth the constant movement of the stars. Towards the end of the 18th century already the Paris Academy of Sciences for instance, decided to no longer accept patent applications for such machines.

[C25] Introducing a hypothetical ether"-, p. 43

The term "ether" has its unique history—as noted above—beginning with Aristotle (384–322 B.C.) and leading all the way to Albert Einstein (1879–1955). In between these two historical figures, one postulated during the 17th century that ether is a medium in which light propagates, understood mainly as a motion of waves. Einstein, in his theory of special relativity, in 1905, banished this ether from physics. According to him light propagates in vacuum. In 1920 he gave the old concept a new meaning, when he postulated that one might attribute physical properties to empty space. Therefore, it would be premature to deny the existence of ether in general. Unfortunately, the 'ether' was corrupted by the followers of a 'German Physics', who had been stirring up trouble as of 1923. In 1934, Heisenberg spoke at the conference of the 'Society of German Natural Scientists and Physicians' where he addressed "The Changes in the Foundations of Exact Natural Science in Recent Time". There he stressed the fundamental significance of the Theory of Relativity and did not hesitate to emphasize Einstein's name. Abroad, it was judged a courageous step, yet inside Germany it brought the dangerous accusation 'White Jew'with it. Heisenberg's 1934 paper was recently republished, in the volume 'Quanten 5' by Konrad Kleinknecht, the president of the *Heisenberg Society* (Stuttgart 2017).

[C26] "The discovery of matter-waves by de Broglie." p. 44

In the years 1923/24, the French physicist, Louis de Broglie (1892–1987), suggested applying the 'wave-particle-duality', familiar from the (immaterial) light, to matter even, for instance, to electrons, in spite of their proven mass. He interpreted them as matter-waves, which they actually are as well. Since, in reality, they do act as waves, they can be used in electron microscopes. In 1929 de Broglie was honored with the Nobel Prize in physics. Erwin Schrödinger (1887–1961) readily took the proposition of electrons as matter-waves and, based on it, formulated in 1926 a wave-mechanics, which culminates in the famous Schrödinger equation. Thus, there existed a second form of quantum mechanics, which stands side by side with Heisenberg's matrix mechanics from the year 1925.

[C27] "Einstein was the first person courageous enough." p. 48

The theory of special relativity by Einstein—like quantum mechanics—is considered today among the great scientific accomplishments of mankind. But, during the 'Third Reich', staunch representatives of a 'German Physics' repudiated it as 'Jewish Physics'. Heisenberg got into trouble when he publicly stressed what he repeats here in private, namely, that one can "not really doubt anymore" the validity of Einstein's conclusions. The theory of relativity, nevertheless, brought with it a problem, even for people like Heisenberg. In his autobiography "Physics and Beyond"(Ref. [1], p. 29) he addresses it in the following way. "I feel kind of betrayed by the logic used for the mathematical framework [of the theory of relativity]. I have understood the theory in my head, but not yet in my heart." One must assume that the politicians during the 'Third Reich' did not understand the theories, neither with their heads nor with their hearts.

[C28] "This 'general theory of relativity' is probably ..." p. 49

A number of experimental investigations over time have confirmed the general theory of relativity. In 1916 Einstein predicted the existence of gravitational waves on account of his theory of gravitation. It refers to waves in the cosmic time-space which occur as a result of accelerated masses. Recently they were verified, awarding the investigating scientists the Nobel Prize in Physics for 2017.

[C29] "The universe has only existed for five billion years" p. 49

According to current knowledge, space has existed for more than 13 billion years, which does not make it any easier to contend with the accompanying "comet's tail of zeros", in Thomas Mann's words.

[C30] "... the lowest temperature" - p. 50

Here Heisenberg is talking about the third law of Thermodynamics, formulated in 1906 by Walter Nernst (1864–1941). Briefly stated, it holds that it is impossible to reach the absolute zero point i.e. the lowest limit of temperature. It has to exist because volumes cannot be negative. The first law of thermodynamics holds that the energy of the world is constant (indestructible); the second expresses that thermodynamic processes are not reversible, which is based on the observation that heat never moves from a colder object to a warmer one.

[C31] "A view of the world that deems the atom itself ..." p. 50

"A view of the world that deems the atom itself a small cosmos"—here Heisenberg refers to the model of the atom, presented in 1913 by Niels Bohr. Many people, to this day, have that as a first image when hearing the word atom. However, it has been for some time already "no longer believable", as one can read and take seriously, and as especially Bohr himself understood.

[C32] "Nothing more than the irregular state of atoms in motion" – p. 52

The first efforts by physicists to understand heat led to the suggestion of a heat substance, named *caloricum*, as previously mentioned. The British physicist Benjamin Thompson (1753–1814) also known as Count Rumford, has gone on to suggest, in his deliberations on heat, when it is generated through mechanical friction, that heat must be understood as a form of motion, as Heisenberg describes it. Count Rumford, though, had to leave open as to what it was, that started to move.

[C33] "Canonical totality" - p. 53

The concept of a canonical totality stems from the American, Josiah Willard Gibbs (1839–1905), who, in 1901, introduced a mathematical distribution function. It allows the description of a system which is in a thermal equilibrium with its surrounding. This is also called canonical ensemble. The word 'canonical', is derived from the Latin '*canonicus*', meaning 'according to rule'.

[C34] "Only in one aspect" – p. 53

The experiment which, according to Heisenberg, necessitated a fundamental change in the physics of the 19th century and rendered the old-world view obsolete, was that so-called 'black body radiation'. When heated, black bodies begin to send out light which changes color with rising temperatures. As Max Planck (1858–1947), around the year 1900, attempted to calculate the distribution of frequencies i.e. the colors of the emitted light, as temperatures rose, he was only successful, after he decided, in a "fit of desperation", to introduce what became popular as the quantization of energy. Thus Quantum Theory was born in a year when Heisenberg was not even yet on this earth.

[C35] "Bohr's theory solves the problem" – p. 55

Actually, it is the Niels Bohr model of atoms, mentioned previously (p. 75), where a nucleus is circled by electrons. Since the circling electrons, in terms of physics, are charges, accelerated in their movement, they must give off energy, making it impossible to hold their positions. To stabilize his model, Bohr, ad hoc, introduced the quantum conditions, at the time new and unusual. It prohibited electrons from continuously shedding their energy. They could, however, exercise quantum leaps, either spontaneously or caused by external influence. Now the model was stable, but still highly unsatisfactory, as mentioned previously.

[C36] "A specifically defined range of possibilities", p. 56

We mentioned above that in philosophical terms it is possible to understand quantum mechanics as the first theory that considers side by side a 'res extensa' (reality) and a 'res potentia', as a realm of possibility, just as Heisenberg had always maintained. The world presents itself as 'potentia' from which, by observation, an actual reality is gained. Accordingly, the world would not be that which actually is the case, as Ludwig Wittgenstein opines in his "Tractatus". However, the world is rather all that it can be, as Anton Zeilinger [16], (*1947) described in his book, lifting 'Einstein's Veil'. Heisenberg completely shared this view. In his lecture on 'Language and Reality in Modern Physics' he writes explicitly, that physicists got used to view "electron orbits and similar terms not as a reality but a kind of 'potentia' in the sense of Aristotelian philosophy". One may add that the Greek put being and possibility side by side, and presumed that it would require a force to change the possible into reality. For this force they introduced the term 'Energeia' and hence energy, which plays a prominent role, not just in physics.

[C37] "The Greeks mapped the lands…" p. 67

Alexander the Great (356–323 B.C.) had Aristotle as his teacher who postulated the existence of two inhabitable regions on earth. He called them 'Oikumene'. One of them reached to India, a term used for the end of the world which, however, nobody had yet seen at that time. When Alexander started out his conquest expeditions in that direction, he had to traverse Mesopotamia with its two rivers, Euphrates and Tigris. It is noteworthy that, in antiquity, one thought that to know the world meant to comprehend creation.

[C38] "Bohr's thesis may be" p. 67

One can find the Bohr thesis Heisenberg mentions, in the speech he gave in 1937 on 'Biology and Atomic Physics' at the International Congress in Bologna, commemorating Luigi Galvani (1732–1798). Bohr's arguments were first published in May 1938 in the protocol of the meeting and later on in the collected works, "Atomic Physics and Human Knowledge" (Dover Books on Physics 2010). Bohr attempts here to carry over into biology the thought of complementary aspects—such as contradictory and yet simultaneously connected images, as for instance in the wave-particle duality of light—which Heisenberg also mentions in the current text (e.g. p. 110) at times. Bohr writes, "In fact, it is plausible to view the specific biological laws as natural laws, complementary to those suitable for a description of properties of inanimate bodies. There is an analogy here to the complementary relationship between the stabilizing properties of atoms and any such phenomena for whose description a space-time coordination of the individual atomic particles is applicable". And Bohr adds "… that this kind of approach is equidistant to the extreme ideas of mechanism and vitalism" (p. 20).

[C39] "Driesch, for example, uses" p. 68

The word 'entelechy' comes from Aristotle, denoting something that carries its destination within. A butterfly, for instance, is produced by the entelechy of the caterpillar which, in turn, is produced by the entelechy of its genome contained in its cells. Hans Driesch (1867–1941) worked as a developmental biologist who concluded, based on his experiments, that in life there has to exist something besides physical-biological activity which constructs its completeness. He named that factor entelechy, sometimes 'Causality of Totality", e.g. in his book 'Philosophy of the Organic' (Leipzig 1928).

[C40] "Force fields of entelechy"- p. 70

Here Heisenberg formulates a thought, conceivably far reaching, which modern molecular biology ought to be aware of. In order to understand the budding configuration of an organism, it is not sufficient to know the molecules that are active in the cells. When an organism prepares its information for building wings in insects or of the eyes, it must know in advance precisely where inside the future body it will be and what exactly is needed there. During the 1950's the biologist Paul A. Weiss (1898–1989), among others, suggested that the developing organisms are preparing, purposefully, an internal morpho-genetic field which serves as an orientation for cells and their genome. This concept has been propagated in recent times by Rupert Sheldrake (*1942) [17], but is currently ignored by a life-science which is thinking quite traditionally these days. Yet in order to understand how organisms, helped by their genes, manage to establish the multitudes of form—the greatest mystery in bio sciences—it might be necessary to locate 'force fields of entelechy', as Heisenberg proposes, so as to comprehend through them the beginning order of life.

[C41] "For instance the x-ray photography of large protein molecules" p. 73

Beginning in the early 20th century, macromolecules, such as proteins— [Heisenberg still terms them 'egg white –Eiweiss']—have been crystallized. Since 1912, the resulting crystals have been studied by X-rays to reveal their structures. The idea comes from Max von Laue (1879–1960), who was awarded the Nobel Prize in Physics in 1914. The X-ray analysis of structures became very influential in the development of molecular biology, making detection of the double helix of DNA possible, the very substance of the genes.

[C42] "The boundary area in which living begins…", p. 74

"The boundary area in which living beings cannot be differentiated clearly from large molecules", was entered in about 1959 with the investigation of viruses that can attack bacteria and kill them, named bacteriophages, for short, phages. They consist of two kinds of molecules, known as proteins and nucleic acids. The research of phages is considered today the origin of molecular biology, and is owed primarily to a physicist who studied with Bohr as a postdoc. He is Max Delbrück (1906–1981) who, in the 1930's, got to know Bohr's thoughts on 'light and life', and, in 1969, received a Nobel Prize in medicine for his investigation on phages. (See also [18]). The 'establishment of natural laws', which Heisenberg kept expecting, has not been successful to this day. Probably largely due to a concept everybody uses today, 'information' whereas Heisenberg was not yet accustomed to its scientific meaning. We are referring to computer science, or informatics, which began its victorious entry after 1945. (See[19])

[C43] "Features of life processes" – p. 76

It is courageous of Heisenberg to speak of life processes "that surely cannot be explained by the principle of natural selection." As far as consciousness goes, which is the topic he refers to here, there are a host of people ready to discuss its possible evolution and its selective advantage. Many fields of thought are relevant—philosophy, theology, sociology, biology, cognitive science, esotericism, rosicrucianism, and so on. Brain-researchers are particularly interested in the question at which level of complexity in a central nervous system something akin to consciousness appears. This topic remains open-ended and exciting. The thought Heisenberg alluded to is connected to Bohr's idea that "the existence of life ... is a given precondition that cannot be further explained ... like the existence of the Planck constant". The sections after page 111 are extensive and can be found with greater detail in the 9th chapter of Heisenberg's autobiography "Physics and Beyond". For instance, according to him, Bohr stated "The real problem is actually: How can that part of reality which begins with consciousness fit with the one descripted in physics and chemistry?" (p. 160).

[C44] "Life is not an experiment of physics" – p. 79

This thought, attributed to Bohr, is found—although not verbatim—in the essay 'Biology and Atomic Physics' from 1938, published in the volume 'Atomic Physics and Human Insight' (Braunschweig 1985). Heisenberg bestows on Bohr's ideas, cumbersome in their formulation, a linguistic clarity, and he often succeeds, especially in his autobiography 'Physics and Beyond'. Thus, also some of Bohr's most evocative sentences are, in a sense, derived through Heisenberg, as noted above.

[C45] "The studies of Carus" – p. 83

The investigations of Carus refer to the writings of the physician and painter Carl Gustav Carus (1789–1869), a natural philosopher from the Romantic period. In 1846—the time in which also the silver string was struck—he presented a book, titled 'Psyche – On the Developmental History of the Soul.' Carus sketches a model of the soul, built around the romantic notion of the unconscious. Its first sentence claims, "The key to the knowledge of the essence inherent in the conscious life of the soul lies in the region of the unconscious." This unconscious, according to Carus, is of divine nature, appearing in humans in embryonic form as an "absolute unconscious"; later on it splits into a 'general' and a 'partial unconscious'. (More on it in: Henry F. Ellenberger, 'The Discovery of the Unconscious' [18]). The Romantics are searching behind visible nature for the secret of its origin which they view as the foundation of the soul as well. They talk of a 'night aspect of nature', which we mention here, because it played an enormous role for Heisenberg's friend, Wolfgang Pauli (1908–1958). Pauli thought and wrote a great deal about the significance of his dreams, and their influence on the scientific world view; Heisenberg knew this and addressed it in some of his popular writings. (Also [19]). Perhaps one ought to mention here as well that Max Planck also alluded to the unconscious in "Scheinprobleme der Wissenschaft [Illusory Problems in Science]" (1946) [20].

[C46] "The White Rose" – p. 90

A white rose commonly stands as a symbol for purity, as Heisenberg writes, whereas readers in the 21st century will associate the 'white rose' with a student group in the resistance movement against the national socialists. The siblings Hans and Sophie Scholl belonged to it, whose activities were monitored by the Gestapo (Secret Police), beginning in the summer of 1942. In February 1943 they were apprehended and in the same month death sentences were handed down and immediately executed.

[C47] "The mind's hostility toward life" – p. 92

The psychologist and philosopher of life, Ludwig Klages (1872–1956), coined the phrase "the mind's hostility toward life" in his main opus 'The Mind as the Antagonist of the Soul'. It appeared in three volumes between 1929 and 1933. (Reprinted Bonn 1972). The ideas of Friedrich Nietzsche (1844–1900) were a strong influence on him. For Klages, the soul is the purpose of the body.

[C48] "Whether specific spiritual abilities… can be biologically inherited" – p. 93

In 1939, Bohr addressed this question of biologically inheritable specific mental abilities in a lecture dealing with 'Epistemological questions in physics and the human cultures'. (First printing in: Nature, vol. 143, p. 268 (1939). Also published in 'Atomic Physics and Human Insight', Braunschweig 1985, pp. 22–30). Heisenberg stays with this topic until p. 140.

[C49] "Art-work's cognitive value" – p. 100

This term appears in the book from 1935 by Martin Heidegger (1889–1976) 'The Origins of a Work of Art.' The philosopher describes art as a linguistic means to experience truth, juxtaposing it with positive science. More on the relationship between Heisenberg and Heidegger is found in the book 'Reflection on Modern Times' (Munich 1983) by Carl Friedrich von Weizsäcker, who expounds on what the two of them have thought and written 'on beauty and art'. Von Weizsäcker mentions that Heisenberg liked to refer back to the definition of beauty, according to the New Platonian Plotin (205–270) who states "Beauty is the penetration of the eternal glory of the One through the material appearance." Von Weizsäcker adds that Heisenberg after such formulations abruptly stops, "in line with his reluctance to speak of the ultimate experience." (p. 156)

[C50] "The Hegelian study of history" – p. 104

The essence of Hegel's approach to history is embedded in the sentence 'World history is progress in terms of freedom – progress that we must recognize as an inherent necessity' [21]. Hegel's (1770–1831) philosophy of history is considered the classical endpoint of historical philosophy, thus comparable to classical physics, something Heisenberg expounds on.

[C51] "A statesman ..." p. 107

"The statesman, on the other hand, often hardly needs to be a human being". Instead, he very well can be inhuman, as one can conclude; making us aware of the fact that one could not have written a more dangerous sentence during the 'Third Reich'.

[C52] "Plato writes in Phaedrus" – p. 108

In his dialogue 'Phaedrus', Plato presents a theory of the soul. He has Socrates (469–399 B.C.) talk of an immortal soul. It steers its soul vehicle across the heaven's vault, to arrive, eventually, at a place where it can perceive the platonic ideas, such as the idea of beauty. In 'Phaedrus' the soul is said to have harmful impacts also, in addition to its creative ones. Many philosophers, over the centuries, have struggled with the Platonic dialogue. Nicolai Hartmann (1882–1950) has pointed out that Plato describes the perception of ideas as an active accomplishment. It entails a unity based on the act of simultaneously perceiving something akin to physicists' dream of a "great unified theory". To reach the above outcomes, it requires guarding "the creative forces of the soul" of which Plato writes.

[C53] "The night in 1920 on the balcony" – p. 110

In his autobiography 'Physics and Beyond' Heisenberg goes into more detail about the night on the balcony. Here, on page 11, is a description of the "experience" the young Heisenberg was privy to, when a violinist appears up on a balcony in a moon-lit night, and begins to play the first, powerful d-minor accords of Bach's Chaconne. They enable Heisenberg to keep talking about the "central realm" which across time has been accessible, through Plato and Bach (1685–1750), and possibly "also now and in the future".

[C54] "Whoever once circled the flame" p. 110

"Anyone who has ever circled the flame shall remain the orbiter of the flame"—These lines are by Stefan George (1868–1933) whose poem ends "only when his gaze lost sight of it …" 'The star of the alliance' was the title of the collection of poems, published in 1914 [22]. George had assembled around him a circle of disciples which was to remain hermetically sealed off to the outside. The verses cited here impressed all the members of the youth movement deeply who had turned to a life in nature, with Heisenberg playing an active role. At the time of his deeply moving event in the courtyard of a castle in the valley of the Altmühl, he was with a troop under his leadership, on its way to a congregation of other youth groups. Here they were gathering to contemplate "where do we go from here" in the aftermath of the First World War with its destruction of the customary order, which had, up to then, provided a "potent center" to people. It was now lost to a whole generation and had to be redefined by the youth themselves ('Physics and Beyond' (Ref. [1], pp. 9–11). Carl Friedrich von Weizsäcker recounts that Heisenberg "spoke more critically of George than I wanted to hear", but instead "could recite from memory more of his poems than I". (Cited in Ulrich Raulff, 'Circle without a Master – Stefan George's Afterlife', p. 386, Munich 2012).

[C55] "In all ages, people have seen it …" p. 114

Plato describes a "divine madness" in his previously cited dialogue 'Phaedrus' (Phaedr. 224a–245a), which is supposedly a term from Socrates. He equates Eros with madness, and distinguishes between a harmful form and a useful one. Socrates ranks as the highest form of divine madness the erotic drive, together with the prophetic madness, the holy madness of rituals, and the poetic madness. For the soul, it is reminiscent of its prenatal state.

[C56] "Hölderlin or Hugo Wolf" – p. 115

These days, Friedrich Hölderlin (1770–1843) is known as an important German poet, but his contemporaries viewed him as demented and crazy, and he had to undergo involuntary treatment at the University of Tübingen. Hugo Wolf (1860–1903) was an Austrian composer, known for his Mörike-Lieder, who in his younger years, contracted syphilis. He was initially hospitalized in a "private clinic for mental illness" and then transferred into the 'State Asylum of Lower Austria'. Both Wolfgang Amadeus Mozart (1756–1791) and Franz Schubert (1797–1828) have lived a much too brief life, unfortunately, which is well known and remains tragic. At the end of Goethe's "Faust", there are the following verses:

> Everything temporary
> Is merely metaphor;
> The inadequate
> Here is event.
> The indescribable
> Here it is done;
> The ever-female
> Is drawing us in.

The musical bars which Heisenberg copies in his handwriting, are the Arietta theme from the second and last movement of the last piano sonata, (op. 111 written in c-minor) by Beethoven. Heisenberg copied it one octave higher. There are two reasons for mentioning it in connection to Goethe's 'Faust'. For one, it demonstrates that it is always the early 19th century with its romantic phases which get Heisenberg's attention. Secondly, the bars call attention to the fact that Beethoven's last sonata has only two movements (not four). This question is discussed in detail in a novel Thomas Mann (1875–1955) begins to write in his exile in California, barely twelve months after Heisenberg, in his Bavarian home, wrote 'Reality and its Order'. That the novel is titled 'Doctor Faustus' needs no mention, nor should it be surprising to anyone. The motive of the Arietta theme consists of three notes d-g-g, rhythmically speaking, like 'Lie-bes-leid' [sorrow of love] or 'Wie-sen-grund' [deep meadow] as the novel does, or as "Hei-sen-berg" if one is permitted to suggest it here. With these notes the composer takes leave, not just from this sonata but from the sonata in general, as a style, a traditional art form, ... "it has fulfilled its fate, reached its goal, beyond which there is no way" as Thomas Mann writes in chapter VIII (without convincing every expert in the history of musicology). It is plausible that Heisenberg too, by pointing to the Arietta theme, is thinking of a leave-taking, a "never come back". Anybody can easily imagine the totality of what he gave up to live in his Germany, or all that will have to be relinquished before the brown nightmare will come to an end, not yet in sight at the time of his writing. Heisenberg searches for and finds comfort at this time in the philosopher Plato's thought, who deems love to be man's yearning for immortality. Specifically, there is one place in the 'Symposium', where Socrates is giving an account of what he learned from the clairvoyant Diotima. According to her, erotic love represents a longing for beauty, is aspiring to immortality and therefore will procreate—as Heisenberg has done in the difficult years of the 'Third Reich'.

[C57] "The next step ... is doubt." p. 116

The French author and philosopher Voltaire (1694–1778), from his perspective as a believer in Enlightenment, wrote his famous statement in 1769. It reads correctly as "If there were no God, one would have to invent him". It stands as Voltaire's counter argument to the anonymous text 'The Three Frauds', referring to Moses, Jesus, and Mohammed. That text is a critique of religion, and was published, for instance in 1994, in the Philosophical Library of the Publishing house, Meiner (Editor Winfried Schröder).

[C58] "A time of a great passion" p. 118

The connection between the time of a great passion and a particularly good year for wine, is made in the fourth section of Goethe's 'Notes and a Treatise for a Better Understanding of the 'West- Eastern Divan'.

[C59] "If you don't want to be my brother" p. 119

"And if you don't want to be my brother, I will bash in your head" is a taunt picked up from the French revolution. Bernhard von Bülow (1849–1929), the Reich-Chancellor under Wilhelm II used it in 1903 during a speech at the Reichstag, thereby helping it gain new popularity.

[C60] "The powerful figure ...is not important" p. 120

Heisenberg refers here to Beethoven's Opera, 'Fidelio', where a prison guard, named Rocco, smuggles an occasional piece of bread to the prisoner, Florestan. A central passage of the opera features this state prisoner, Florestan, in the darkness of a dungeon, as he wakes up. He begins intoning an aria with the words "Oh, those were good days", which we may point out here, refers to the fact that the silver string back then was not yet torn. Two things are noteworthy: Heisenberg often refers to Beethoven, who throughout his life has been thinking politically. Also, what Maestro Wilhelm Furtwängler (1886–1954) said, who conducted the first Fidelio performance after National Socialism had come to an end. The scene on stage featured a concentration camp and barbed wire. He remarked that Beethoven's music "never before has appeared as grand, never so vitally important as just now" (cited in Eleonore Büning, "Let us talk about Beethoven", Salzburg 2018). Heisenberg was already aware of it in his core, when he wrote his 'Reality and its Order'.

[C61] "The question what reality really is" – p. 121

The fairy tale Heisenberg is talking about is called 'The Little Shepherd Boy', found in the 2nd edition of the Brothers' Grimm 'Children's and House Fairytales', published in 1819. In this fairytale, on p. 152 of the collection, the king is summoning a shepherd boy and asks him three questions. The first one: How many drops of water are in the world oceans? And the little boy answers, he would tell the king the number as soon as he could stop all the tributaries to the ocean, so that not one droplet could be added anymore. The second question asks: How many stars are in the sky? And the shepherd boy draws innumerable little dots on a paper, each one for a star. And the third question asks: How many seconds are in eternity? The answer can be read in Heisenberg. The king is satisfied with all responses. And if neither of them have died, then they are still alive today.

References

1. Heisenberg W (1971) Physics and beyond, encounters and conversations. Harper, New York
2. Huizinga J (1955) Homo ludens. Beacon Press, Boston
3. Schiller F (2004) On the aesthetic education of man. Dover Books on Western Philosophy
4. Heisenberg W (1958) The physicist's conception of nature. Hutchinson & Co, London

5. Fischer EP (2015) Werner Heisenberg – a wanderer between two worlds. Springer, Heidelberg
6. Zimmer E (1936) The revolution in physics. Faber and Faber, London
7. von Uexküll JJ. Umwelt und Innenwelt der Tiere [Environment and the inner world of animals]. Since 2014 it has been available in an edition with commentary. Springer, Heidelberg
8. Crahan D (2018) Helmholtz – a life in science. University of Chicago Press, Chicago
9. Piaget J (1978) Das Weltbild des Kindes [Theory of cognitive development], German edn. DTV, Stuttgart
10. Malebranche N (1997) The search after truth. Cambridge University Press, Cambridge
11. von Weizsäcker CF (1992) Zeit und Wissen [Time and knowledge]. Hanser, Munich, p 801
12. Lorenz K (1941) Kants Lehre vom Apriorischen im Lichte gegenwärtiger Biologie [Kant's theory of the a priori in light of current biology]. Blätter für Deutsche Philosophie [J Ger Philos] 15:94–125
13. Lorenz K (1978) Behind the mirror – a search for a natural history of human knowledge. Mariner Books
14. Heisenberg W (1959) Physics and philosophy. Harper New York. [Also under the title: *Form and Thought* that appeared in the "Yearbook of the Bavarian Academy of Arts", vol. 6 (Munich, 1960)]
15. Eilenberger W (2018) Zeit der Zauberer [Time of the magicians]. Clett-Kotta, Stuttgart
16. Zeillinger A (2010) Dance of the photons. Farrar, Straus and Giroux, New York
17. Sheldrake R (2003) The sense of being stared at. Arrow, New York
18. Ellenberger H (1981) The discovery of the unconscious. Basic Books, New York
19. Fischer E-P (2004) The glimmering night view. Lengwil, Switzerland
20. Planck M (1946) Scheinprobleme der Wissenschaft [Illusory problems in science]. Barth, Leipzig
21. Quoted in Emil Angehrn (1991) Geschichtsphilosophie [Philosophy of history]. Kohlhammer, Stuttgart, p 90
22. The works of Stefan George (1974) University of North Carolina Press
23. Fischer EP, Lipson C (1985) Thinking about science – Max Delbrück and the origins of molecular biology. Norton, New York
24. Fischer E-P (2010) Information: Kurze Geschichte in Fünf Kapitel. Jacoby & Stuart, Berlin